Wright Brothers: Then and Now

Daniel E. Cleary

Cleary Creative Photography, Ltd.

4140 Linden Avenue, Suite 210

Dayton, Ohio 45432

www.ClearyCreativePhoto.com

Dan@ClearyCreativePhoto.com

ISBN 978-0-578-83665-2

Library of Congress Control Number 2021900554

First printing edition 2021

V4

Table of Contents

Dedication

To my Dad and Father-In-Law
Frank W. Cleary and Paul J. Stamas

I wish you were here to see what you inspired me to do.

"The past didn't go anywhere."
Utah Phillips

Forward

People have asked me how I became interested in photography. I inherited my passion for this art form from my father, Frank William Cleary, although it took me many years to clearly see the connections. He was born in Cincinnati, Ohio, in 1919. After graduating from Purcell Catholic High School, he took up photography. He cleaned out the old coal room in the family home and converted that space into a dark room where he processed black and white film and made prints. He attended college part-time and also worked for the Wright Aeronautical Corporation as a male secretary. In 1940, the United States reinstated the military draft, and in 1941, Frank was drafted into the Army. He was eligible for a deferment because his company made airplane engines for the Army Air Corps, but as he said, "I had the fever and thought I had a duty to the country. Besides, it was only for a year." He went to Fort Bragg, North Carolina for basic training and took his camera with him.

The attack on Pearl Harbor on December 7, 1941, changed his enlistment timeline from one to four years. After basic training, he remained at Fort Bragg, where he worked as a secretary in the Colonel's office. In the fall of 1942, he said "yes" to an offer to attend Officers' Candidate School and was off to Fort Sill, Oklahoma.

One day, shortly before leaving for Fort Sill, he laid his camera down and left the barracks for a few minutes. When he came back, the camera was gone.

In April 1943, Frank was commissioned as a 2nd Lieutenant in the U.S. Army and transferred to the 128th Armored Artillery Battalion at Fort Cooke, California. His division was sent to Europe in 1944 and assigned to General Patton's 3rd Army. He landed at Utah Beach just over a month after D-Day. No one in his division had yet been to war. His first assignment was as an Artillery Forward Observer in a tank. My Dad had no "forward observing" training. He didn't know what to do and was scared. His tank was the first in line on the day they prepared to move into battle.

He stood up with the top hatch open, watching below him as General Patton walked up to the Division Commander and said, "Your job is to take Brest," and walked away. The fighting was intense, but my father returned unharmed.

Late in August 1944, Frank was transferred to the 212th Artillery Battalion to become an Air Forward Scout. The 212th had taken many casualties, and the man he was replacing had died in the line of duty. As an adult, he once told me, "At night in my prayers, I would ask, 'Lord, what are you saving me for?' That is why I would volunteer for missions. I knew I was saving someone else's life because I knew I'd be OK."

Forward

As an Air Forward Scout, he was part of an eight-man crew: two pilots, two observers, a sergeant, and three enlisted men. They were ahead of the Army and on their own. They would go up in the air in the morning and call-in artillery strikes on enemy positions. While the Army was taking new ground, the forward observers sometimes had a day off in the town they had just occupied. The Army would collect all the radios, guns, and cameras. Somehow, he managed to get a camera with film, set up a darkroom inside a tent, and in the dead of night, processed film and made prints, all in the middle of World War ll. Much later in life, my Dad wrote a memoir of his time in the Army. It includes many of the photos he took and developed himself.

On December 11, 1944, five days before the Battle of the Bulge, his division received heavy shelling. He and his pilot, Lou Blumberg, jumped into their plane and went up looking for targets. They could approximate the enemy shells' general direction, but every time they would get within visual range, the shelling would stop. Lou and my Dad decided to keep flying lower and lower over the enemy until someone fired on them. Finally, a German soldier fired his rifle at them. They pulled up and called in the artillery strike. He and his pilot received the Silver Star

for Valor. In October 1945, he was honorably discharged from the Army and returned to Cincinnati, where he met and married my mother, and they started our family.

My father never talked about the war when I was a child, so I was surprised at his response when in high school, I told him about a photography class I was thinking of taking at The Living Arts Center. It was a facility organized by the City of Dayton and Dayton Public Schools, explicitly designed for education in the arts. He immediately said that I could use his camera from the Army. I said, "What camera?" He then said I could use his enlarger in the attic if I wanted, and I said, "What enlarger?" It was the first time I had any idea that he had an interest in photography. I used his camera and set up a darkroom with his enlarger in my bedroom.

In college, I took math, science, and photography classes. In the end, I decided to stick with photography. I received a Bachelor of Fine Arts from Wright State University and a Master of Fine Arts in Photography from Cranbrook Academy of Art in Bloomfield Hills, Michigan. I returned to Dayton after college, married my wife Maria, raised two children, opened Cleary Creative Photography, and for over 30 years have worked as a professional photographer, doing what I love.

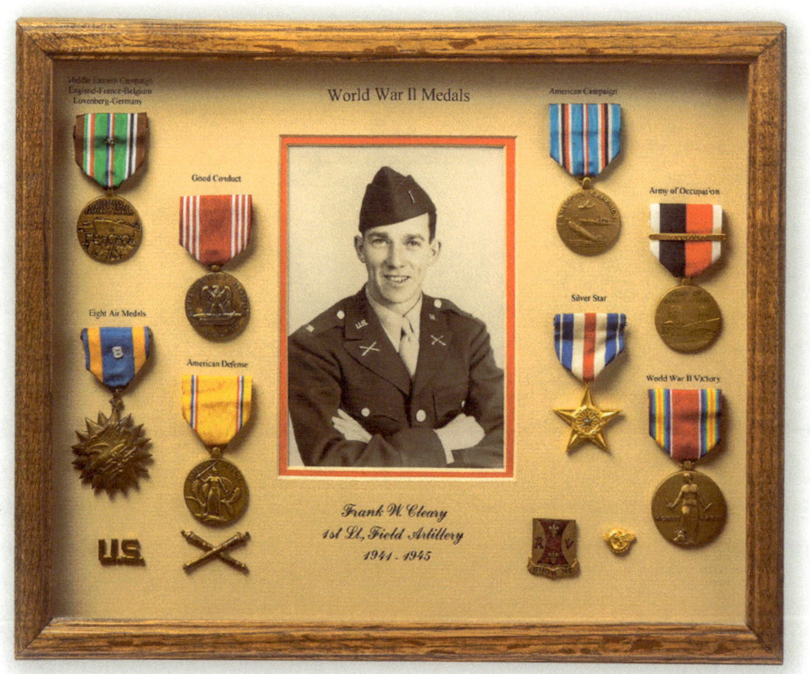

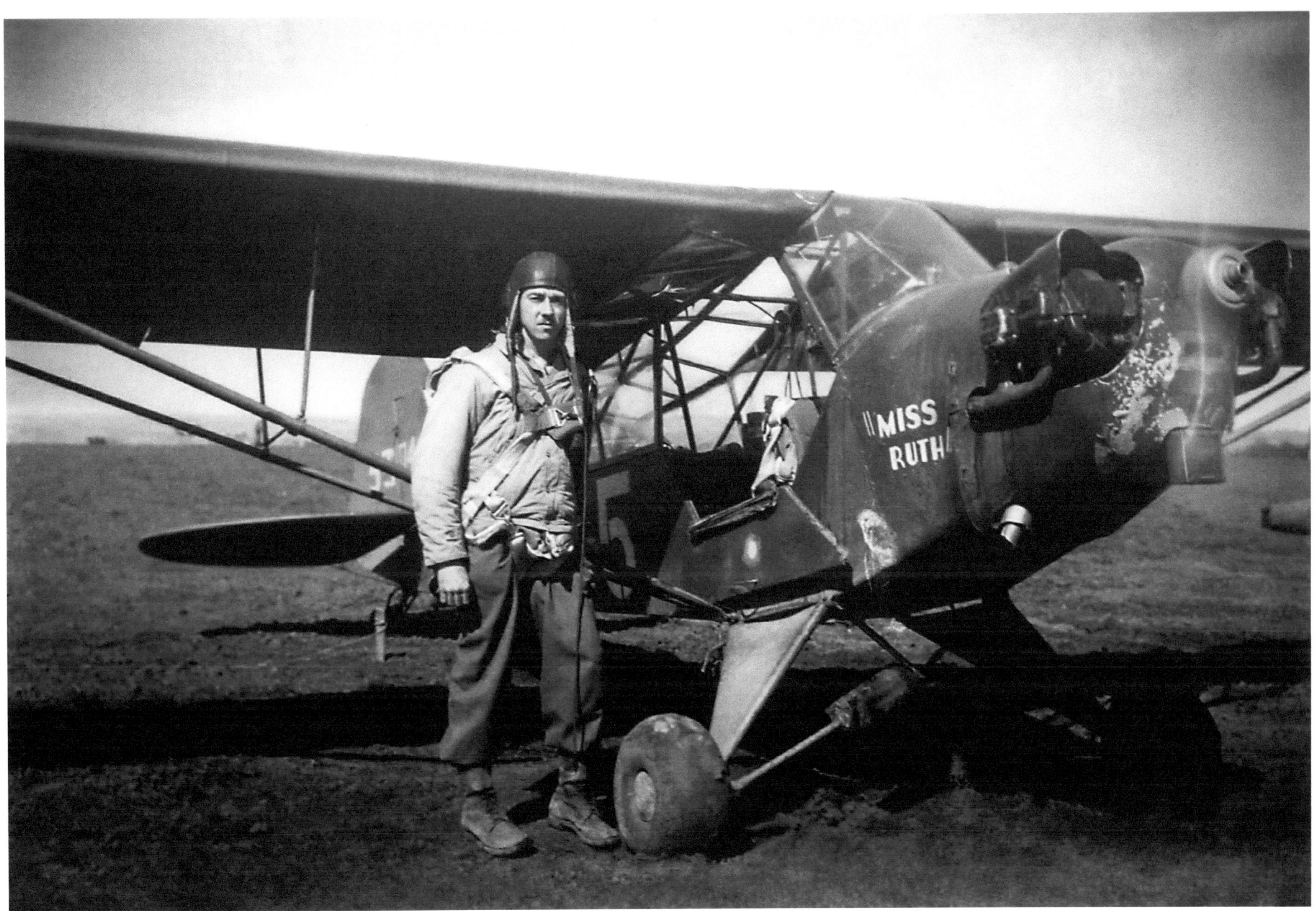

MISS
RUTH

Introduction

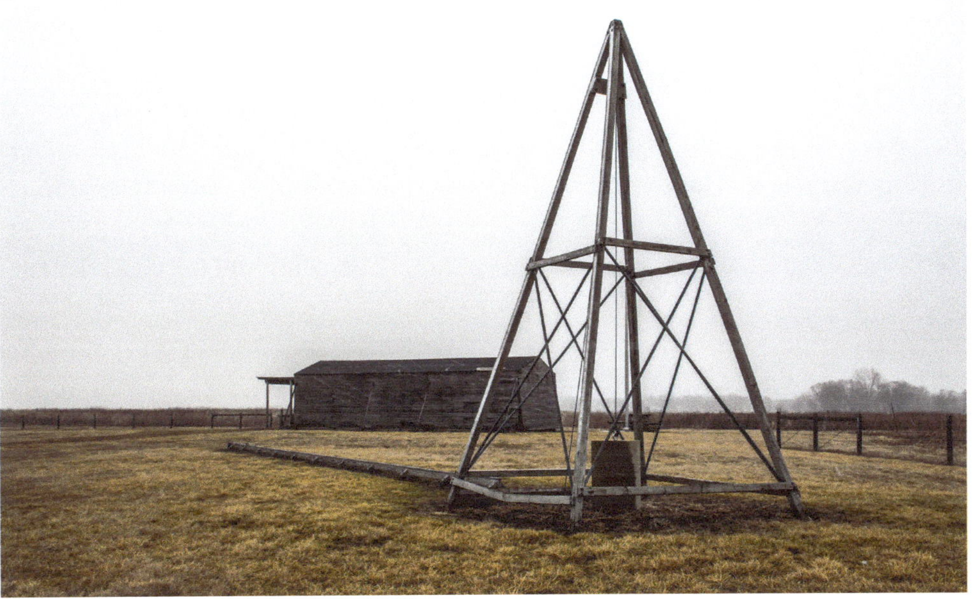

In 2016, I listened to the audio version of David McCullough's book, *The Wright Brothers*, while working in my studio office. Afterward, it occurred to me that I had never visited Huffman Prairie, even though Dayton, Ohio is my hometown and also the home of the Wright Brothers. So, on a cold and slightly snowy day in February, I took my camera and visited the site where the brothers perfected flight. Standing alone in the field that day, I felt the past and the present converging, and I could almost see and hear the Wright Brothers' plane above me. David McCullough's biography was so vivid that the story of Amos Root witnessing the 1904 flyer circling Huffman Prairie popped into my head. The experience transformed me. Although I only took a few exposures that day, and I didn't know it at the time, this vision planted within me the seeds for this series of photographs.

Wright Brothers: Then and Now is about the past mingling with the present. The Wright Brothers were interested in photography and even sold cameras in their bicycle shops. They used photography in their process of discovery. Through research, I found many original historic images taken by the Wright Brothers and others, whose locations I could still access. I began traveling to places like Kitty Hawk, North Carolina, Washington, DC, Detroit, Michigan, New York City, Le Mans and Pau, France, as well as locations in Dayton to create my photographs. I then digitally combine my images with the historical pictures in Photoshop, blurring the divisions between then and now and creating a world where past events and modern-day activity blend. I include a written narrative within each photograph that gives context and life to the image and the people behind the history.

The Bike Shop

The Dayton Aviation Heritage National Historical Park in Dayton, Ohio, is part of the National Park Service. There you can visit the fourth of Orville and Wilbur's bike shops. The bike shop was two blocks from their home on Hawthorn Street. They worked in this shop from the spring of 1895 to the fall of 1897.

While there, the Wrights started manufacturing a line of bicycles and purchased state-of-the-art equipment to build them. By the fall of 1896, they were actively studying the idea of powered flight. By making their line of bicycles, they became integrally familiar with this equipment and its capabilities. They would later use that knowledge to build their gliders and the first airplane.

I went to the bike shop at the National Park and created a series of interior photographs that I felt would match the historical image. In the photo, Wilbur is working on a lathe. The actual lathe he was working on in the old photograph is on display at the park. Many of the tools also on display and scattered on the workbench are the Wrights' original implements. I could almost feel Wilbur still working in this room. I merged the contemporary and the historic by extending the floor from my image throughout the entire photograph. The effect is that Wilbur could be standing on the floor, working right now in the space.

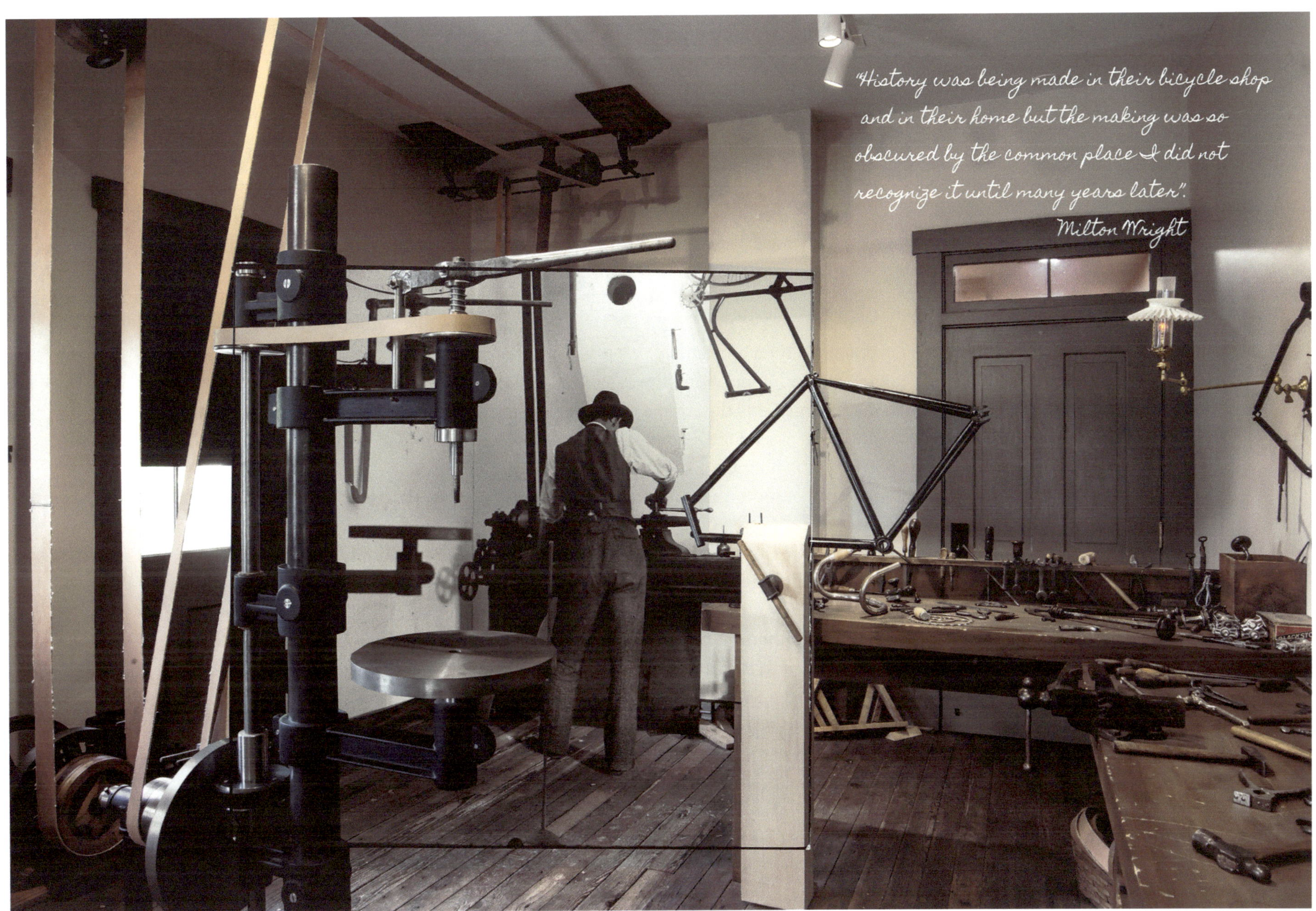
"History was being made in their bicycle shop and in their home but the making was so obscured by the common place I did not recognize it until many years later".
Milton Wright

7 Hawthorn Street

"If I were to give a young man advice as to how he might succeed in life. I would say to him, pick out a good Father and Mother and begin life in Ohio."
Wilbur Wright

Orville Wright was interested in photography and made this historical photograph around 1897 of Daniel Henderson on the sidewalk in front of the Wright house at 7 Hawthorn Street. When I saw this image, I imagined Orville taking his camera out on a beautiful day to ensure the camera was in good working order.

Along comes Daniel Henderson, and after seeing the camera on a tripod, he asks Orville, "Hey Mr. Wright, what ya doin'?" This imagined encounter was the source of my inspiration for this image. It made me look at the Wright Brothers as photographers instead of two men just using photography to document their trials

Orville could have composed the photograph with Daniel in the center of the frame, but instead, he placed Daniel in the lower right-hand corner of the image, directly in the lines of perspective. For me, this is a much more sophisticated composition.

I was familiar with this quote from Wilbur, and I liked the photograph of Daniel. I immediately put both together in my mind. All I had to do was create a current day photograph of the house to combine the elements. But where was the house? Henry Ford acquired the Wright Brothers' home and cycle shop and relocated the buildings from Dayton, Ohio, to his Dearborn, Michigan museum complex in 1937. What now remains at 7 Hawthorn Street in Dayton is a replica of the front porch and house façade, which was enough to create this image.

7 Hawthorn Street may be one of the more complex photographs in the series. The complexity is in how all the parts of the two images merge. The wrought iron fences in both pictures come together. The porch railings go from old to new. The sidewalk Daniel is standing on goes from new to old and back to new again. The area behind Daniel is from the contemporary photograph, but your first thought is it must be old. The City of Dayton put up a replica streetlamp that I included inside the old photograph frame, which you see as an original element. I even cut out in between the old fencing areas so the modern photograph would show through. By the time I had finished, this photograph had at least 75 Photoshop layers.

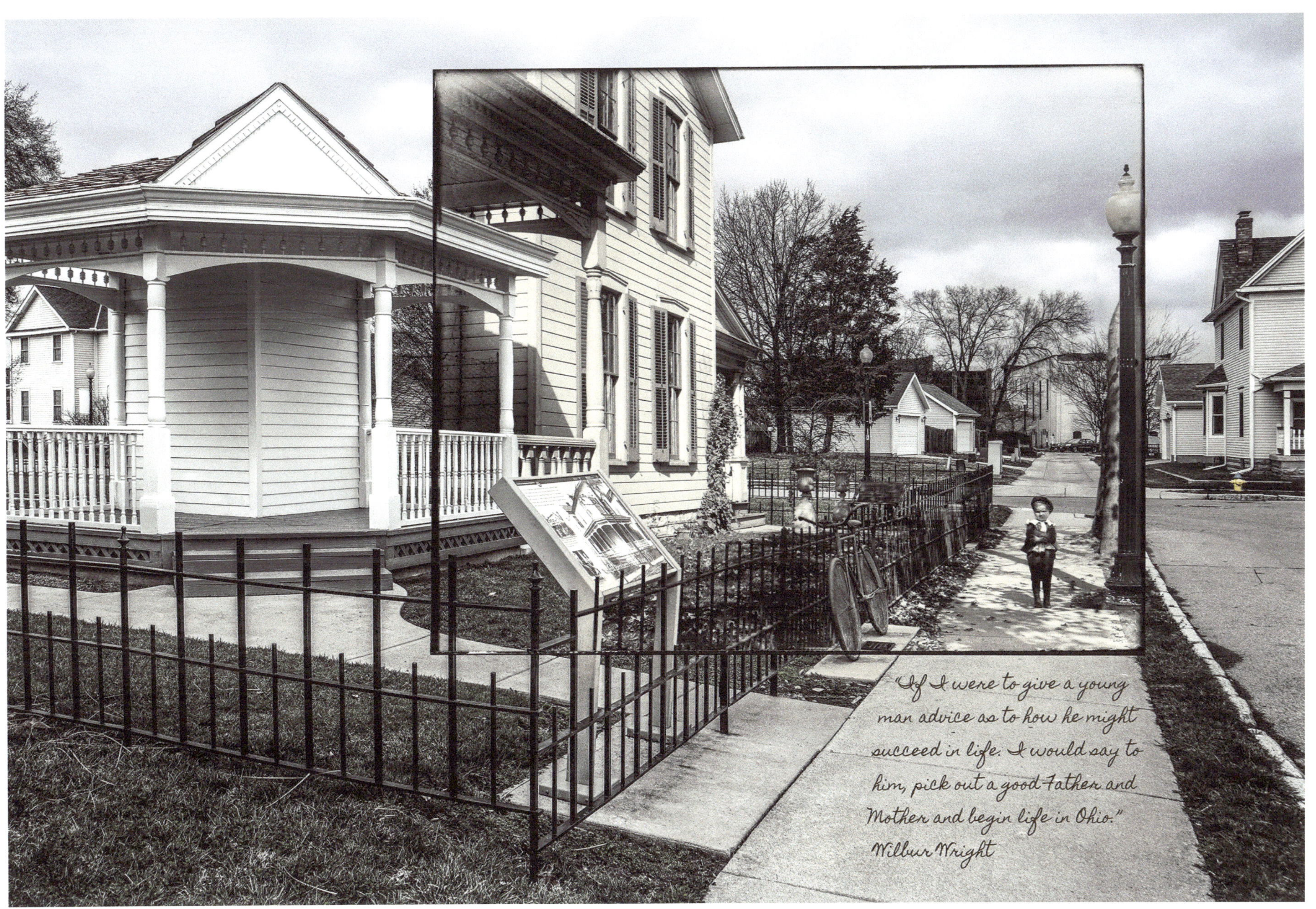
If I were to give a young
man advice as to how he might
succeed in life. I would say to
him, pick out a good Father and
Mother and begin life in Ohio."
Wilbur Wright

Kite Flying

The Wright Brothers tested their ideas about flight at Kill Devil Hills, just south of Kitty Hawk, North Carolina, part of the Outer Banks. Wilbur wrote to his father, Milton, to explain his reasons for choosing the area: "I chose Kitty Hawk because it seemed the place which most clearly met the required conditions…At Kitty Hawk, which is on the narrow bar separating the Sound from the Ocean, there are neither hills nor trees, so that it offers a safe place for practice. Also, the wind there is stronger than any place near home and is almost constant." In September of 1900, the brothers arrived in Kitty Hawk and set up camp to begin their kite experiments.

In the summer of 2016, my wife Maria and I drove to the Wright Brothers National Memorial in Kill Devil Hills, North Carolina. As we were exploring the site, Park Rangers started to pass out kites to visitors. It was the perfect opportunity to photograph, especially knowing the Wright Brothers used kites in their testing. I view this photograph as a form of street photography, and it's meant to be fun. There was quite a bit of moving people and kites around in Photoshop to make a good composition. I think the quote works well with this image and I like Orville's sense of humor.

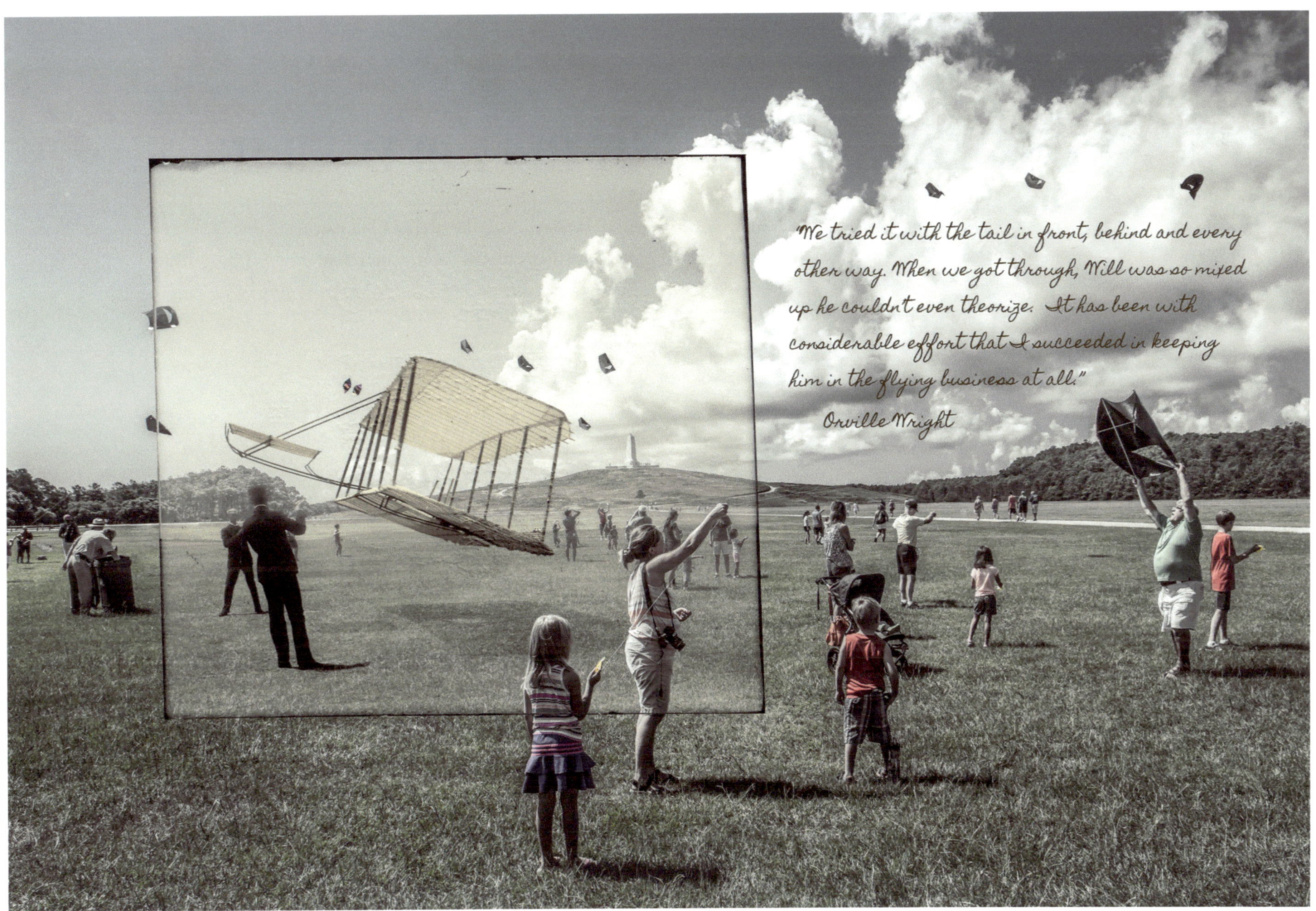

"We tried it with the tail in front, behind and every other way. When we got through, Will was so mixed up he couldn't even theorize. It has been with considerable effort that I succeeded in keeping him in the flying business at all."
Orville Wright

View From The Top Of Big Kill Devil Hill

"For some years, I have been afflicted with the belief that flight is possible to man. My disease has increased in severity, and I feel that it will soon cost me an increased amount of money if not my life. I have been trying to arrange my affairs in such a way that I can devote my entire time for a few months to experiment in this field."
Wilbur Wright

I created this image from the top of Big Kill Devil Hill at the Wright Brothers National Monument. At that moment, I felt as if I were standing in the exact spot from which the historic photograph was taken. I lined up my horizon line with the horizon of the Wright Brothers' historical image in Photoshop. The Wright Brothers' barn's location in the foreground was almost identical in the historic and modern pictures.

Octave Chanute was an American civil engineer and aviation pioneer. He provided the Wright Brothers with advice and helped them to publicize their flying experiments. Wilbur Wright read Chanute's book, Progress in Flying Machines, and wrote to him. Chanute quickly saw the genius of the two brothers. He and Wilbur wrote many letters to each other. Two assistants accompanied Chanute to North Carolina to work with the Wrights in 1902. The brothers were making 600-foot flights in their fully controllable glider. The airplane was finally ready for an engine. By the following year, they took off and flew under power.

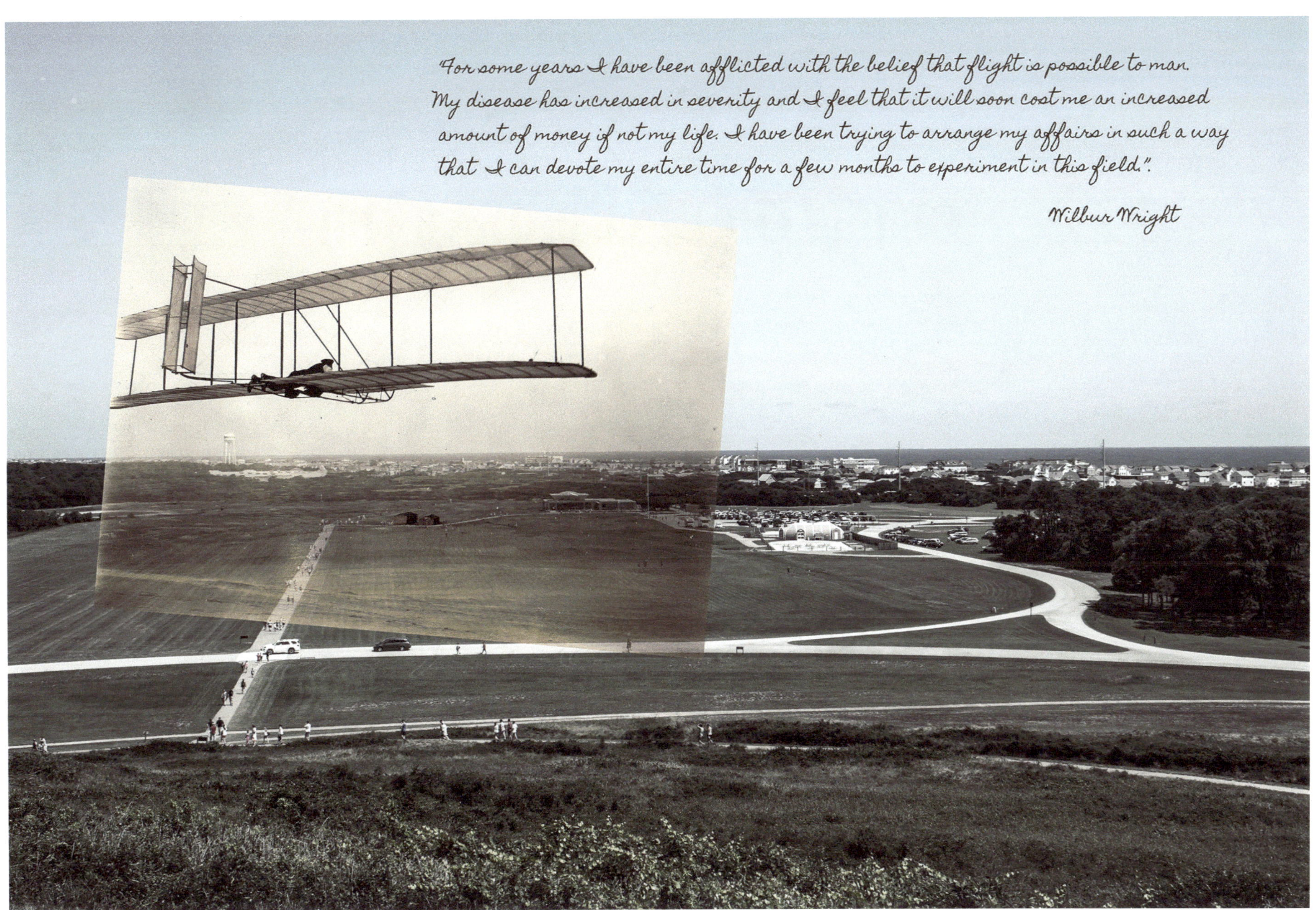
"For some years I have been afflicted with the belief that flight is possible to man. My disease has increased in severity and I feel that it will soon cost me an increased amount of money if not my life. I have been trying to arrange my affairs in such a way that I can devote my entire time for a few months to experiment in this field."

Wilbur Wright

Gliding Off Big Kill Devil Hill

"It has occurred to me that you would get still flatter glides by making sure that the center of gravity coincides exactly with the center of pressure, and possible by decreasing the angle of your rudder to 4 degrees. Please take plenty of snapshots."
Octave Chanute

This image looks up Big Kill Devil Hill to the Wright Brothers Monument at the top. The monument and National Park are actually in Kill Devil Hills, North Carolina. At the time of the Wright Brothers' first flight in 1903, the town did not yet exist. It did not receive its municipal charter until 1953. Kitty Hawk, popularly noted as the famous first flight site, is approximately four miles north and was the nearest settlement at the time. The Wright Brothers excelled at using photography to document their flight experiments. Wilbur Wright and Octave Chanute frequently exchanged letters. The "Please take plenty of snapshots" statement at the end of one of Chanute's letters resonated with me.

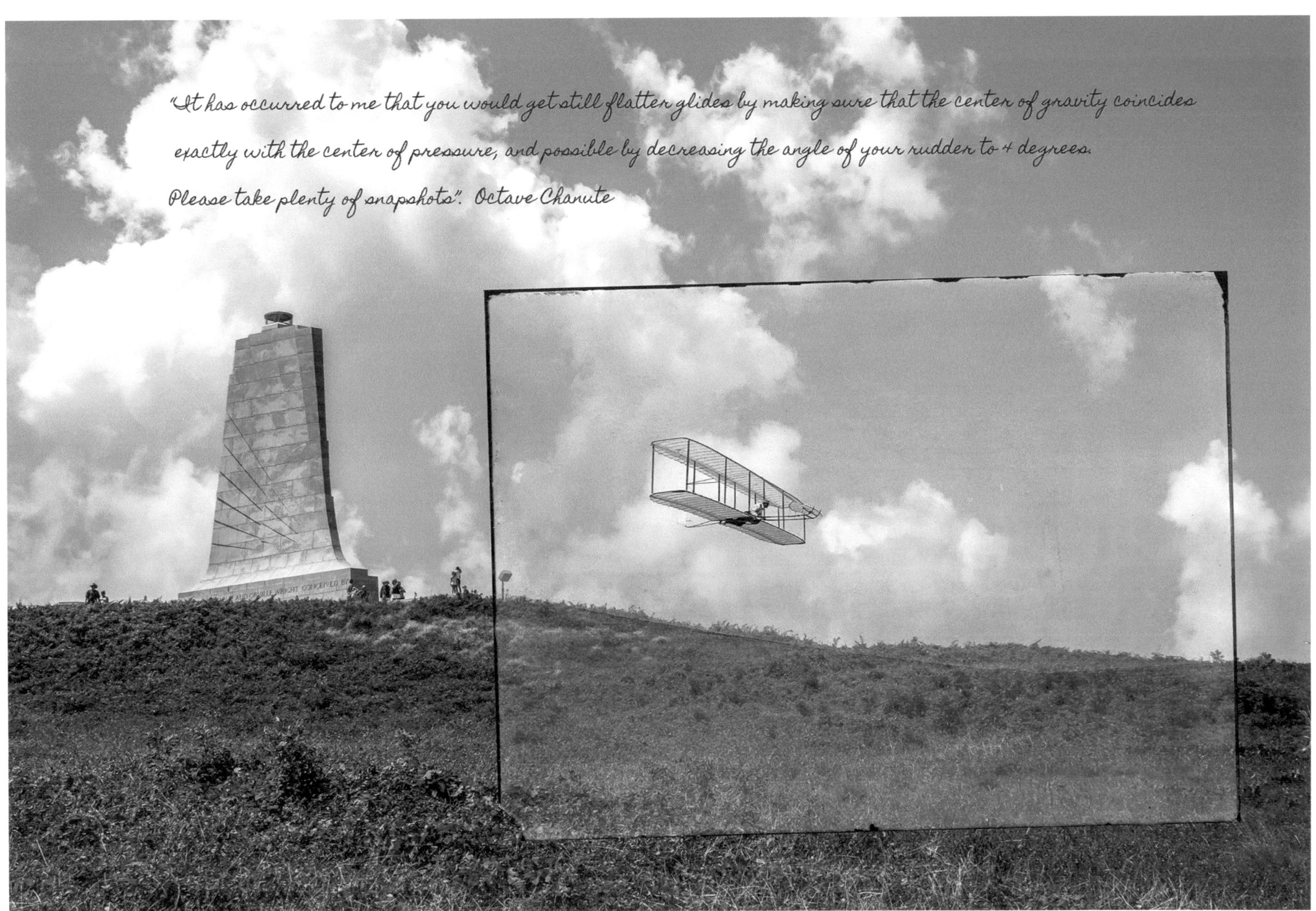
"It has occurred to me that you would get still flatter glides by making sure that the center of gravity coincides exactly with the center of pressure, and possible by decreasing the angle of your rudder to 4 degrees. Please take plenty of snapshots". Octave Chanute

Waiting To Fly

"Wind of 6 to 7 meters blowing from west and northwest in morning. We completed repairs by noon and got the machine out on the tracks in front of the building ready for a trial from the level. The wind was gradually dying and by the time we were ready was blowing only about 4 to 5 meters per sec. After waiting several hours to see whether it would breeze up again, we took the machine back in."
Orville Wright

This historical photograph for *Waiting To Fly* was taken on November 24, 1903. The Wright Brothers had been working hard during the prior months to finish the flying machine, but bad weather and mechanical problems held them up. The quote used in this photograph is from Orville's diary on December 16th, and it states they took the machine out in the morning because the weather reports looked good to make a test flight. I call this "the day that almost was the first flight." I can imagine there was a sense of excitement, thinking this might be the day of the first flight. It took a few hours to get prepared for a test flight, and by the time they were ready with the plane, the wind was calm. They had a plane ready to go but no wind.

I created my photograph at the Wright Brothers National Memorial in 2016, 113 years after the original flight. On the Memorial grounds, there is a replica of the camp the Wright Brothers used. My goal was to integrate the contemporary park visitors with Wilbur, utilizing the building as an anchor. I moved a few people in and around the barn. I even left park rangers inside the barn, so it looks as if they are standing next to Wilbur. I can imagine Wilbur looking out and thinking, "Who are all these people when we have so much work to do?"

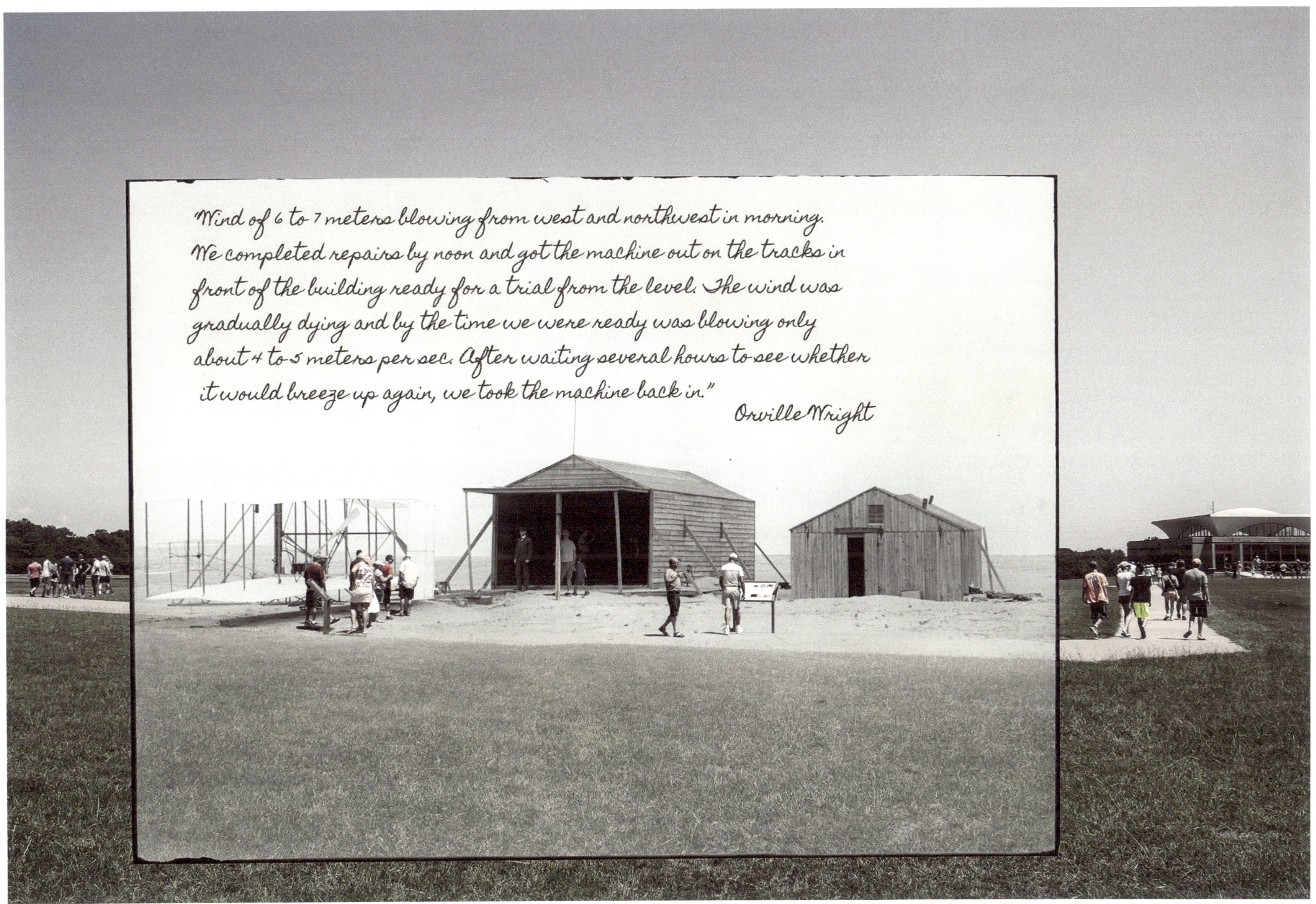

"Wind of 6 to 7 meters blowing from west and northwest in morning. We completed repairs by noon and got the machine out on the tracks in front of the building ready for a trial from the level. The wind was gradually dying and by the time we were ready was blowing only about 4 to 5 meters per sec. After waiting several hours to see whether it would breeze up again, we took the machine back in."
Orville Wright

First Flight

"I got on the machine at 10:35 for the first trial. The wind, according to our anemometers at this time, was blowing a little over 20 miles. The machine lifted from the track just as it was entering the fourth rail. A sudden dart when a little over a hundred feet from the end of the track or a little over 120 feet from the point at which it rose into the air, ended the flight."
Orville Wright

John T. Daniels took the famous photograph of the Wright Brothers' first flight. Daniels was a member of the U.S. Life-Saving Station in Kill Devil Hills, North Carolina, and knew little about photography. His job was to squeeze a rubber bulb, forcing air through a tube that pushed the camera lens's shutter release. Daniels didn't remember squeezing the bulb. I am very familiar with this type of shutter release system, and I can imagine that when he saw the plane take off, he instinctively tightened his hand around the bulb. The Wrights knew that the shutter had been released but didn't know if they had a photograph until months later when they processed the film in their darkroom at their home at 7 Hawthorn Street in Dayton, Ohio.

My main goal in this body of work is to fuse past and present images. As I looked through my photographs for a photo to pair with this historical image, a gentleman in the crowd caught my attention. He had the same stance as Wilbur. Look for him in the far-right corner of the photo. For me, this visual pairing helps to integrate the two images taken 113 years apart.

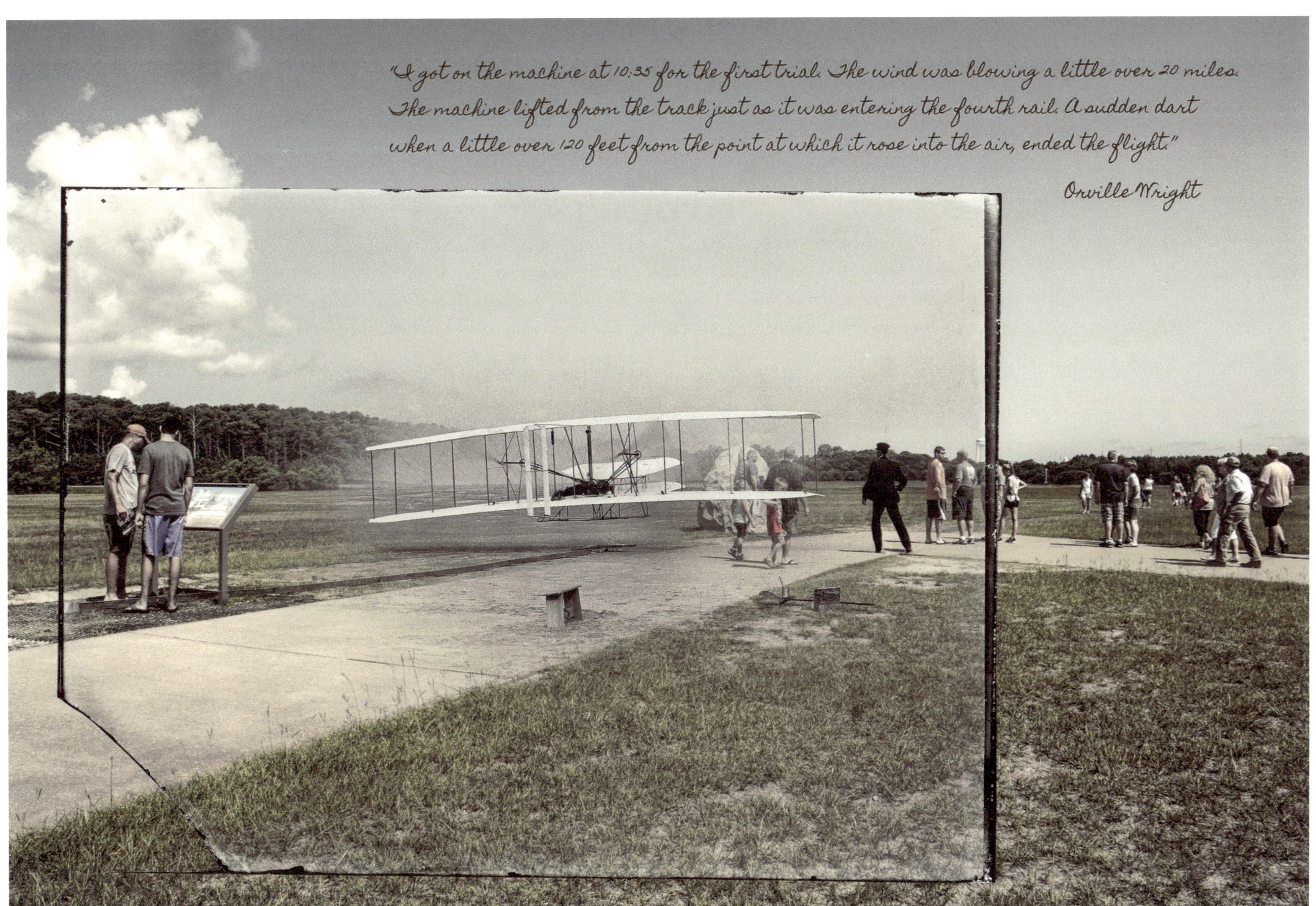
"I got on the machine at 10:35 for the first trial. The wind was blowing a little over 20 miles. The machine lifted from the track just as it was entering the fourth rail. A sudden dart when a little over 120 feet from the point at which it rose into the air, ended the flight."

Orville Wright

Brothers

"From the time we were little children, my brother Orville and myself lived together, played together and worked together, and in fact thought together. We talked over our thoughts and our aspirations so that nearly everything that was done in our lives has been the results of conversations, suggestions, and discussions between us."
Wilber Wright

I discovered this historical photograph in the Library of Congress collection and knew I wanted it to be part of my series. I think of it as the Wright Brothers' first public relations photograph showcasing their machine. Wilbur and Orville stood next to the 1904 *Wright Flyer II* outside of the barn at Huffman Prairie, where it was stored when not in use. The brothers made 105 flights at Huffman Prairie during that year, totaling 49 minutes in the air.

A reproduction of the barn is located at Huffman Prairie Flying Field National Historical Park next to Wright-Patterson Air Force Base near Fairborn, Ohio. I merged the historic photograph's original structure with the reproduction barn in my current-day image in this photograph.

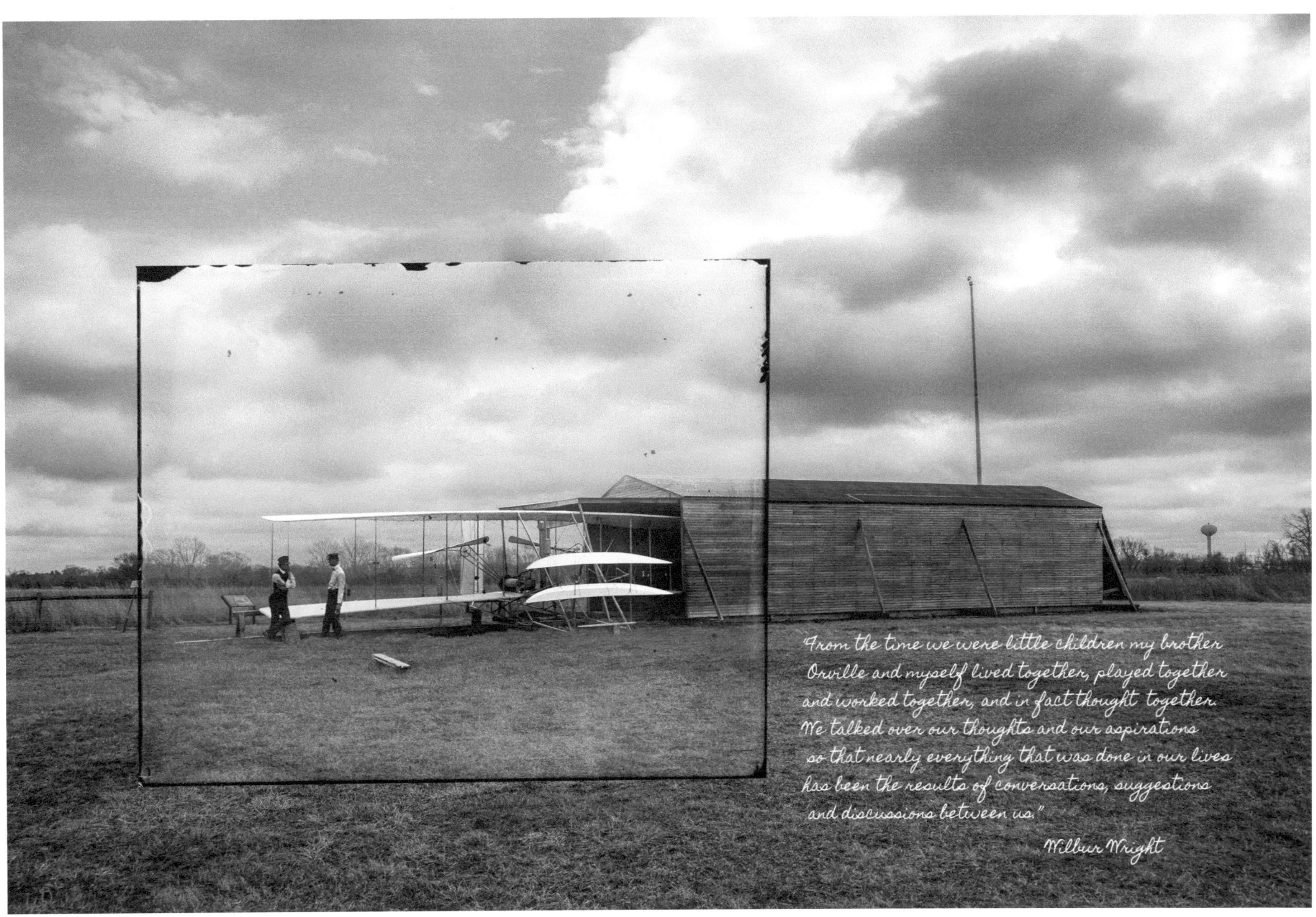
"From the time we were little children my brother
Orville and myself lived together, played together
and worked together, and in fact thought together.
We talked over our thoughts and our aspirations
so that nearly everything that was done in our lives
has been the results of conversations, suggestions
and discussions between us."
Wilbur Wright

Amos Root

"When it turned that circle and came near the starting point, I was right in front of it, and I said then, and I believe still, it was the greatest sight of my life. Imagine a locomotive that left its tracks and climbed up in the air right toward you. A locomotive without any wheels but with wings instead, spread 20 feet each way, coming right towards you with a tremendous flap of its propeller and you have something like what I saw. I tell you friends, the sensation that one feels is something hard to describe."

Amos Root

This photograph is the first image I created in my series. I went to Huffman Prairie in February 2016 after listening to the audio version of David McCullough's biography of the Wright Brothers. As I stood there in snow flurries, I could hear Amos Root's words in my head. I could see the Wright Brothers' plane flying in my mind. I started researching photographs the Wrights had taken at Huffman Prairie and realized the landscape hadn't changed in 100 plus years.

After the Wright Brothers success on December 17, 1903, in Kitty Hawk, they wanted to find a suitable location for test flights closer to their home in Dayton. They approached Torrence Huffman, who owned a farm east of Dayton, about using his land to test their machine. He agreed to give them access to the property without charge as long as no harm came to his livestock. Huffman Prairie was about eight miles from downtown Dayton and could be reached by trolley, making travel there and back easy.

In June and July of 1904, the Wright Brothers worked on their flying machine at Huffman Prairie, mainly without witnesses. During this time, Amos Root from Medina, Ohio, owner of a beekeeper's supply business, contacted the brothers. He wrote a newsletter called *The Gleanings in Bee Culture* and was known as the "Bee Man of Ohio." He loved new technology and enjoyed publishing such stories in his newsletter.

Through correspondence, Root and the Wrights struck-up a friendship, finally inviting him to come to Dayton. On Tuesday, September 20, 1904, he arrived at Huffman Prairie to watch Wilbur fly the airplane in a complete circle.

"When it turned that circle, and came near the starting point, I was right in front of it, and I said then, and I beleive still, it was the greatest sight of my life. Imagine a locomotive that left its tracks and climbed up in the air right toward you. A locomotive without any wheels but with wings instead, spread 20 feet each way, coming right towards you with a tremendous flap of its propeller and you have something like what I saw. I tell you friends the sensation that one feels is something hard to describe."
Amos Root

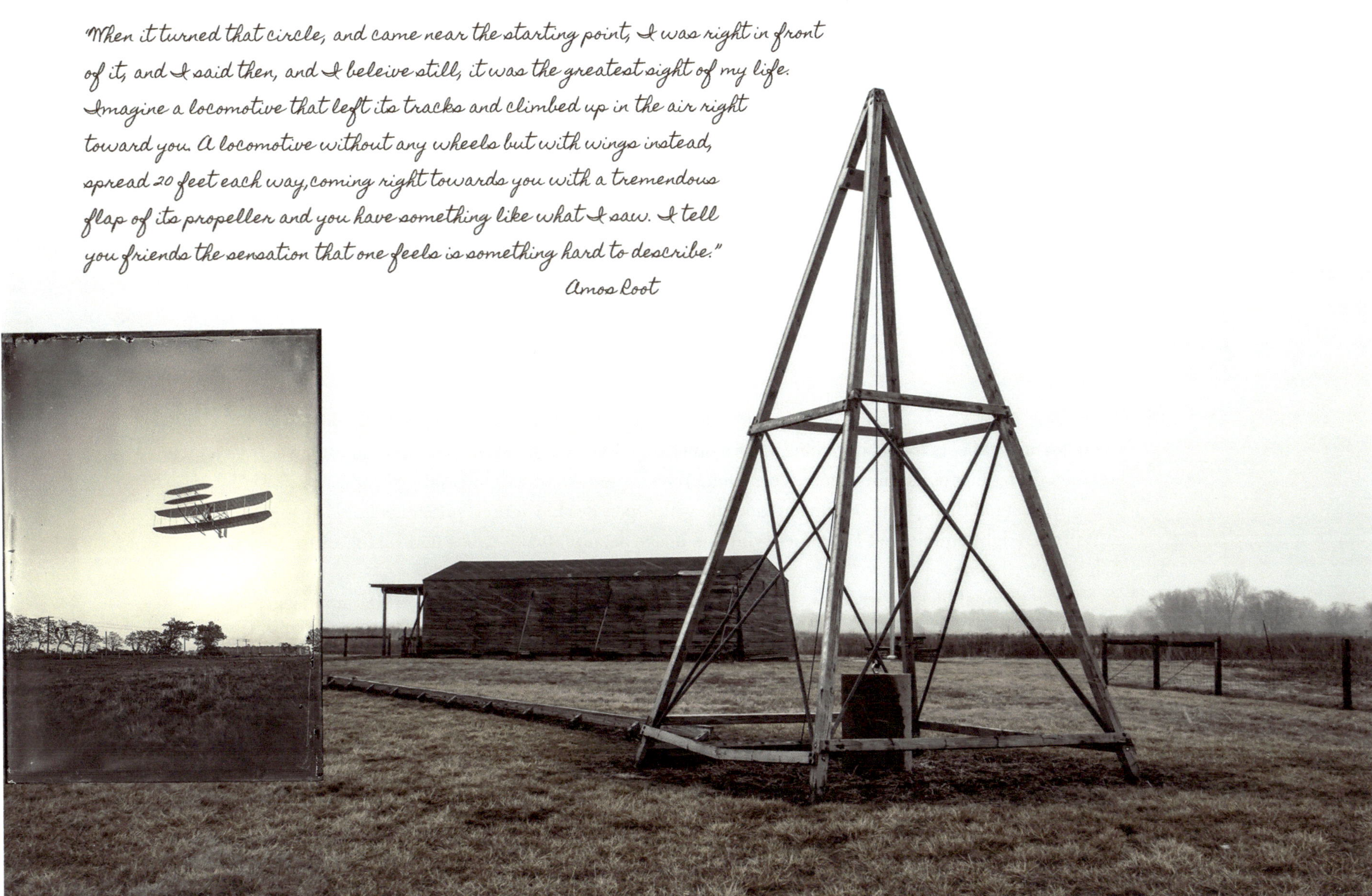

Flight #85

"When you know, after the first few minutes, that the whole mechanism is working perfectly, the sensation is so keenly delightful as to be almost beyond description. Nobody who has not yet experienced it for themselves can realize it."
Wilbur Wright

I consider *Flight #85* the centerpiece of my photographic series. The clarity of the historic photograph reflects the modern landscape's quietness. Even today, when you are at Huffman Prairie, a feeling of calmness comes over you. You can sense the Wright Brothers' presence and easily imagine their plane buzzing around the wide-open field. A retired pilot once told me that Wilbur Wright's quote that I included with this image accurately describes how he felt every time he was in the air.

The original historic image was taken at Huffman Prairie on November 16, 1904, and is so sharp that you can almost see the propeller spinning. On that day, Orville flew a distance of 1,760 feet in 40 seconds, going about 30 miles per hour. The Wrights' original photos were made on glass plates. Often the glass plates cracked, and much of the time, the broken pieces were lost. In this case, both the main image and the fractured part remain. As a photographer, the glass plate's crack is meaningful because it shows process and product simultaneously.

"When you know, after the first few minutes, that the whole mechanism is working perfectly, the sensation is so keenly delightful as to be almost beyond description. Nobody who has not yet experienced it for himself can realize it."
Wilbur Wright

Flight #41

In 1905, the Wright Brothers continued to perfect their flying machine at Huffman Prairie. This historic photograph was taken at Huffman Prairie on September 29, 1905, with Orville flying at an altitude of about 60 feet.

The Wrights had to navigate around a thorny honey locust tree in the middle of Huffman Prairie while making test flights. The day before the historic photograph was taken, Orville came too close to the tree, clipping it with a wing and almost crashing. He regained control by putting the plane into a brief dive and landed safely. Today at Huffman Prairie, there is another tree in the middle of the flying field. I included it in my landscape photograph, paying homage to the honey locust tree that was in the field in 1905 and to Orville's skill in successfully negotiating around that tree.

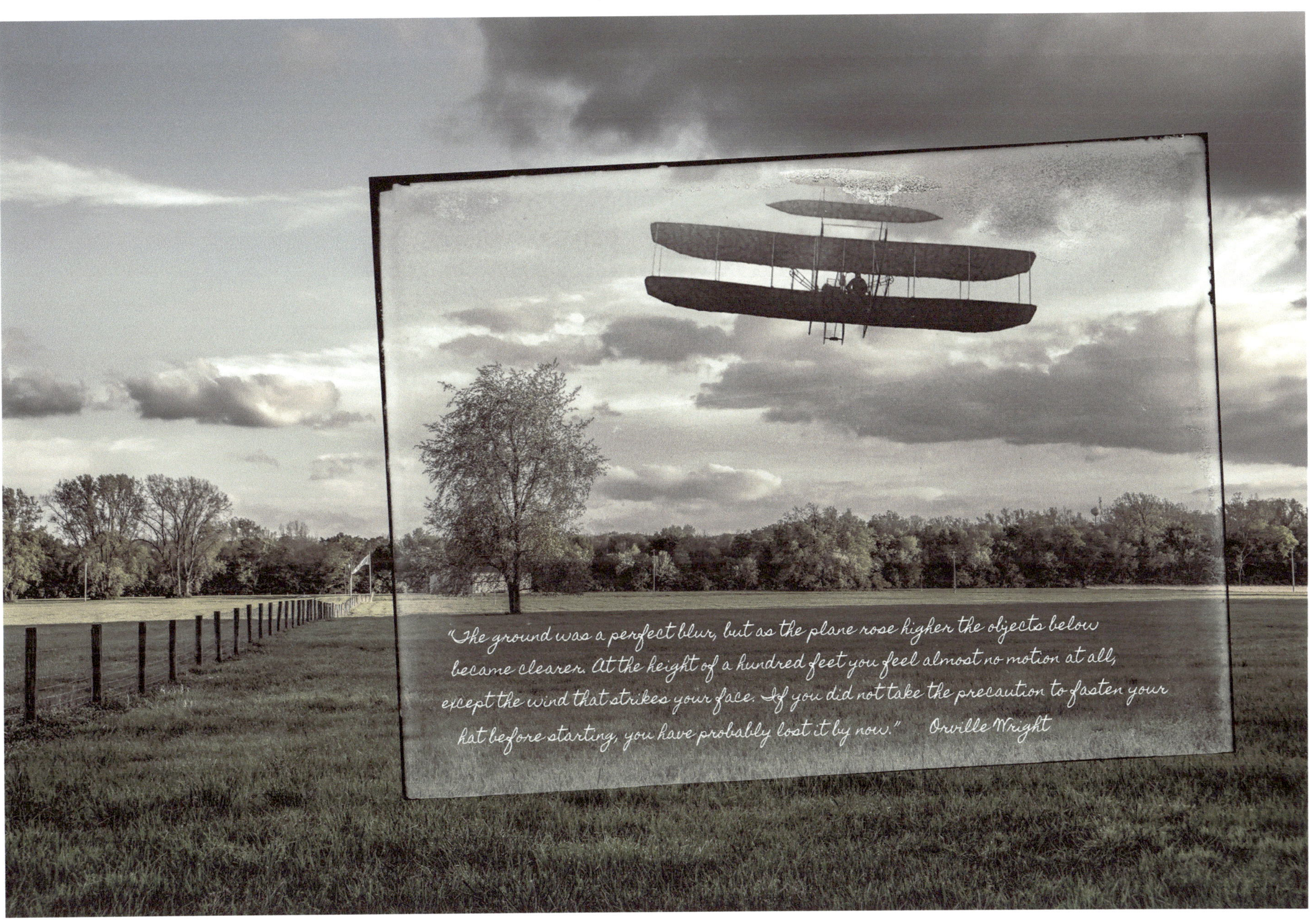
"The ground was a perfect blur, but as the plane rose higher the objects below became clearer. At the height of a hundred feet you feel almost no motion at all, except the wind that strikes your face. If you did not take the precaution to fasten your hat before starting, you have probably lost it by now."   Orville Wright

Traveling With The Wind

"You find yourself facing toward the point from which you started. The objects on the ground now seem to be moving at a much higher speed, though you perceive no changes in the pressure of the wind on your face. You know that you are traveling with the wind."
Orville Wright

I woke up one fall morning to dense fog. Dayton, Ohio, does not have many foggy days, so I decided to go out to Huffman Prairie to take some photographs. I paired the image I made that morning with the historical photograph of Flight #23 taken at Huffman Prairie on September 7, 1905. Orville was at the controls, making two complete circles of the field in 2 minutes and 45 seconds. Standing in the fog, I could imagine Wilbur and Orville using the tree as a target to circle while perfecting flight during their trials at Huffman Prairie.

"You find yourself facing toward the point from which you started.
  The objects on the ground now seem to be moving at a much higher speed,
   though you perceive no changes in the pressure of the wind on your face. You know
that you are traveling with the wind"   Orville Wright

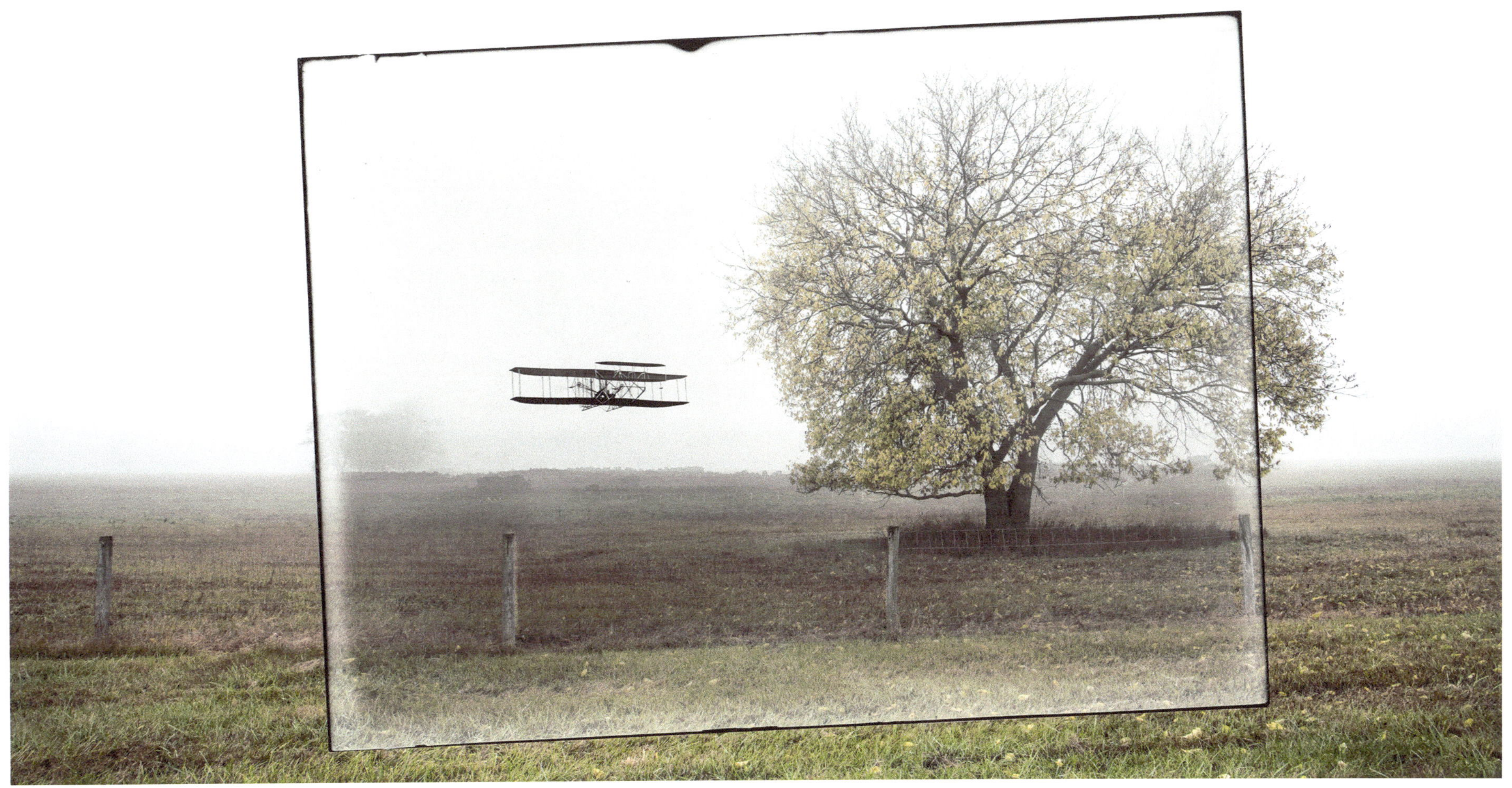

Le Mans Racetrack

"It is not really necessary to look too far into the future. We have seen enough already
to see that it will be magnificent. Only let us hurry and open the roads."
Wilbur Wright

Wilbur Wright gave the first public flying machine demonstration on August 8, 1908, at Les Hunaudières race track in Le Mans, France. Fifteen years after Wilbur made his historic flight, the city of Le Mans became famous for the 24 Hour of Le Mans race. The Le Mans (horse) racetrack where Wilbur flew in 1908 is encircled by the 24 Hours of Le Mans track and is still used for horse races today. The Les Hunaudières flying demonstration finally proved to the world that man could fly and that the Wrights were the ones who made it possible.

In 2019, I traveled to France with my family to create additional images for this series and was lucky to be in France the week of the 24 Hours of Le Mans race. Fridays before the weekend race, the track is open to the public, and I could roam the venue freely with the crowd.

I made this image on the racetrack, looking up the first hill right out of the pit lane. I meshed the historical photo of Wilbur's 1908 flight demonstration with my Le Mans track image. I love that a few spectators in the crowd appear to be pointing their cameras at Wilbur flying overhead.

The innovations in flight that he and his brother initiated are staggering. Their invention paved the way for incredible aviation achievements, from more sophisticated, faster, and larger aircraft to space vehicles capable of interplanetary travel. To commemorate the Wright Brothers' monumental triumphs, fabric and wood pieces from the airplane Kitty Hawk went to the moon on the Apollo 11 lunar module, Eagle, which landed on the moon.

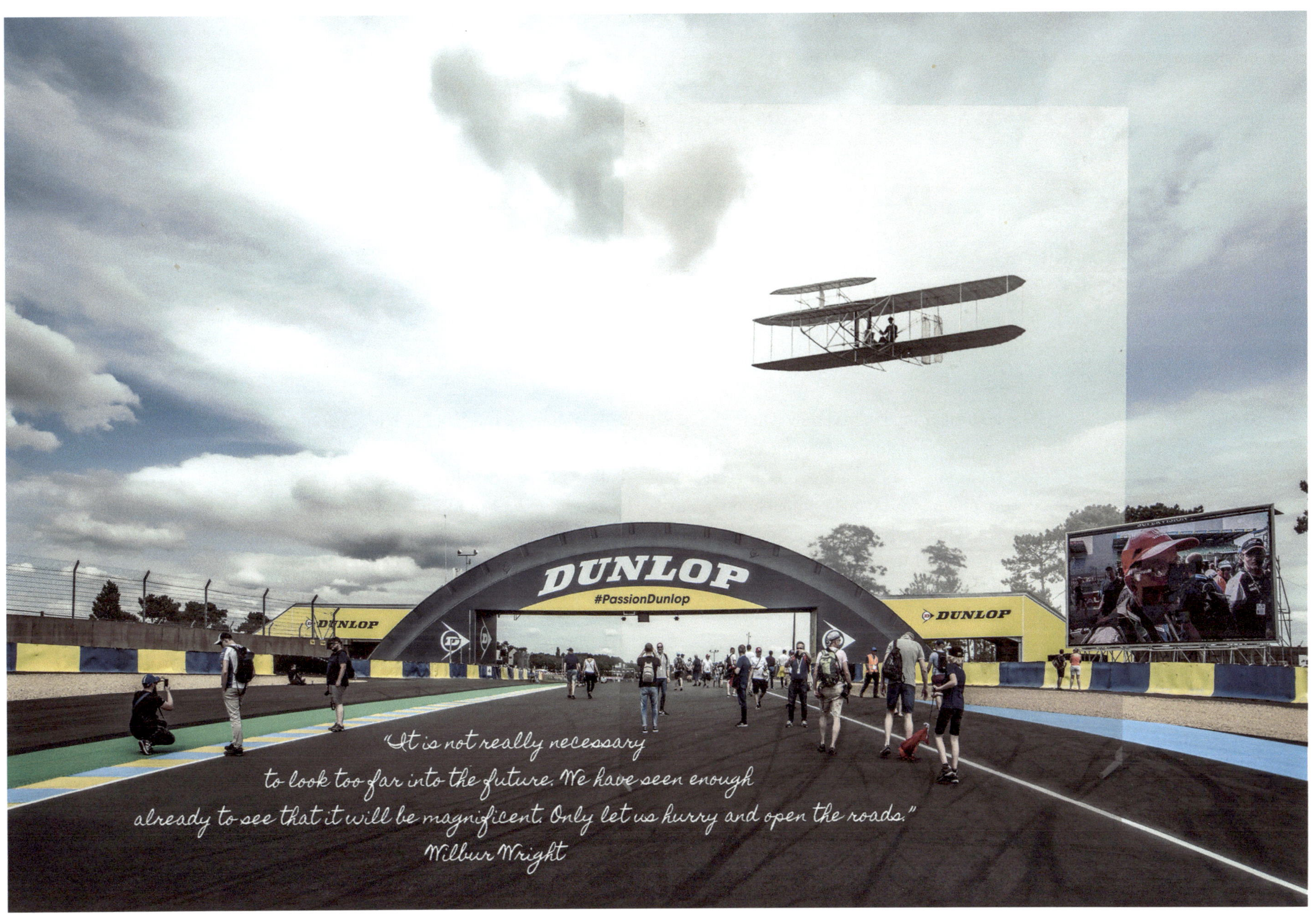

DUNLOP
#PassionDunlop
DUNLOP
DUNLOP
"It is not really necessary
to look too far into the future. We have seen enough
already to see that it will be magnificent. Only let us hurry and open the roads."
Wilbur Wright

King Alfonso XIII Of Spain

The historical image, taken in Pau, France, is of Wilbur shaking the hand of King Alfonso XIII of Spain. Wilbur Wright had become an instant worldwide celebrity with his flight demonstrations in France, yet he maintained a low-profile while there. Visiting the Louvre Museum and taking long walks alone were among his favorite things to do in Paris. The inventor slept in the hanger with his airplane instead of staying in the local hotel in Le Mans, France. He was a private person who, in most photographs, was stoic. I was surprised by the genuine smile and self-confident expression in this image and was thrilled to use it. Digitally I put the King of Spain inside the ropes with the race teams and Wilbur on the other side among the general public.

My modern-day photograph is of the spectators watching the racing crews doing their last-minute fine-tuning of the cars before the start of the 24 Hours of Le Mans race the next day. I was lucky enough to be one of those spectators that year.

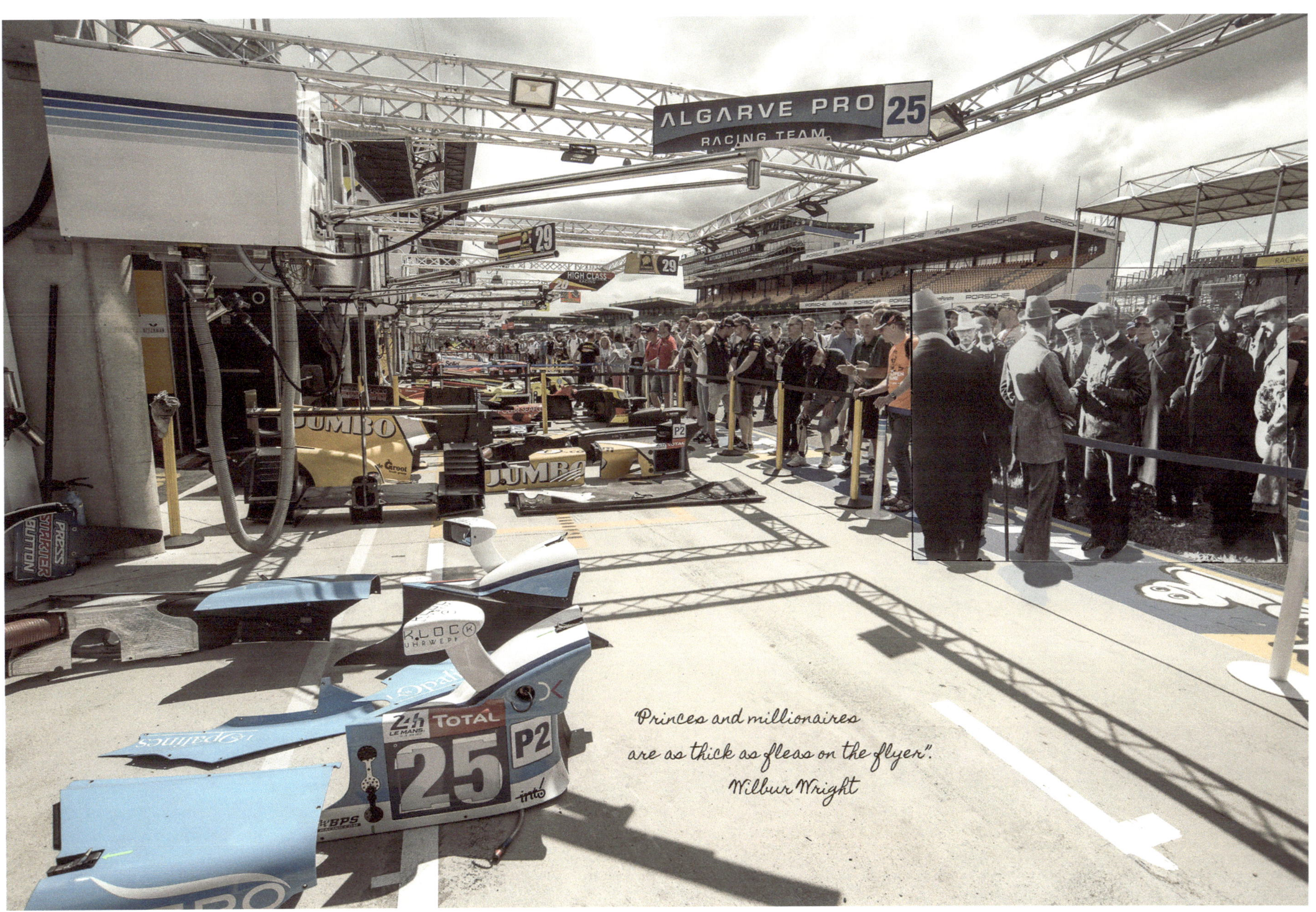

"Princes and millionaires
are as thick as fleas on the flyer".
Wilbur Wright

Speed Test At Fort Myer

*"He flew 67 minutes in a 16 mile wind, handling his plane like a chauffeur, and road the wind
as deliberately as if he were passing over a solid macadam road. Nothing I have ever seen
is comparable in action to this gliding bird."*
Gutzon Borglum

While Wilbur was making successful flights in France, Orville gave flight demonstrations for
the American government in Washington, DC. Up to this point, no one in the United States believed
that these two boys from Ohio could have solved the mystery of flight.

On Thursday, September 3, 1908, Orville made his first test flight for the U.S. Army at Fort
Myer. Fort Myer is located in Arlington, Virginia, just outside of Washington, DC. The base is
surrounded by Arlington National Cemetery and is currently the home base for the military Joint
Chiefs of Staff. His record-making performances at Fort Myer drew praise from the government.
Crowds of officials and noteworthy people were eager to see him fly. The quote for this image comes
from the Mount Rushmore sculptor Gutzon Borglum's reaction to witnessing one of Orville's flights.

I took many photographs as I walked the parade grounds at Fort Myer that day, not knowing
which historical photos I would use. Later, when searching for an image, I came across this photo
of photographers documenting one of Orville's flights at Fort Myer. I immediately identified with
their stances. How many times had I, as a professional photographer, taken that same pose, camera
at the ready, to capture the action in front of me?

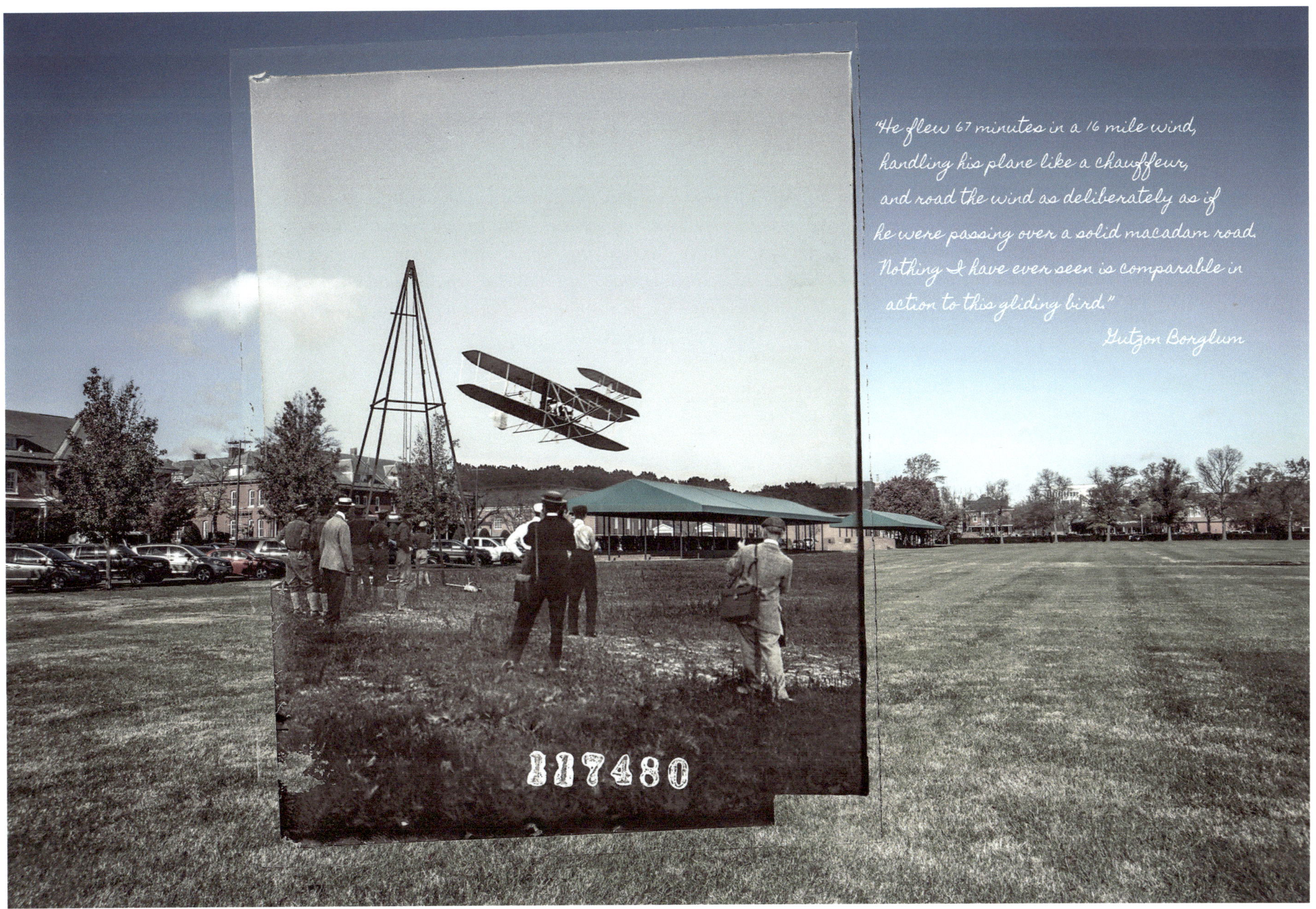

"He flew 67 minutes in a 16 mile wind,
handling his plane like a chauffeur,
and road the wind as deliberately as if
he were passing over a solid macadam road.
Nothing I have ever seen is comparable in
action to this gliding bird."
Gutzon Borglum

The Crash

"As the aeroplane dashed off the rising track, Lieut. Selfridge waved his hand gayly to a group of army officers and newspaper men and threw back some laughing remarks that were drowned out by the whirl of the propellors. As he swept around Selfridge evidently was enjoying himself thoroughly. It was on the fourth and final lap that the propellor blade broke. The aeroplane took a short and sharp dive and crashed into the field. Dust arose in a yellow pall over the great white man-made bird that had dashed to its death."
New York Times Newspaper

On September 17, 1908, Lieutenant Thomas Selfridge flew as a passenger with Orville Wright as he demonstrated a *Wright Flyer* for the U.S. Army at Fort Myer. Selfridge was keenly interested in flight and just a few months earlier, in May of 1908, became the first U.S. military officer to pilot a modern aircraft. Wright and Selfridge had just completed 4 ½ circles of the course when suddenly the propeller broke and sent the *Flyer* into a nose-dive. The plane crashed into the ground, seriously injuring Orville and killing Lieutenant Selfridge. Lieutenant Thomas Selfridge became the first person to die in an airplane crash.

I visited Fort Myer on a warm and sunny day in November of 2018, 110 years after Orville Wright gave flying demonstrations for the U.S. Army. The parade grounds are surprisingly small. I walked around undisturbed for a couple of hours, taking photographs and imagining the crowds of dignitaries watching in awe as Orville flew and witnessing in horror the chaos after the crash.

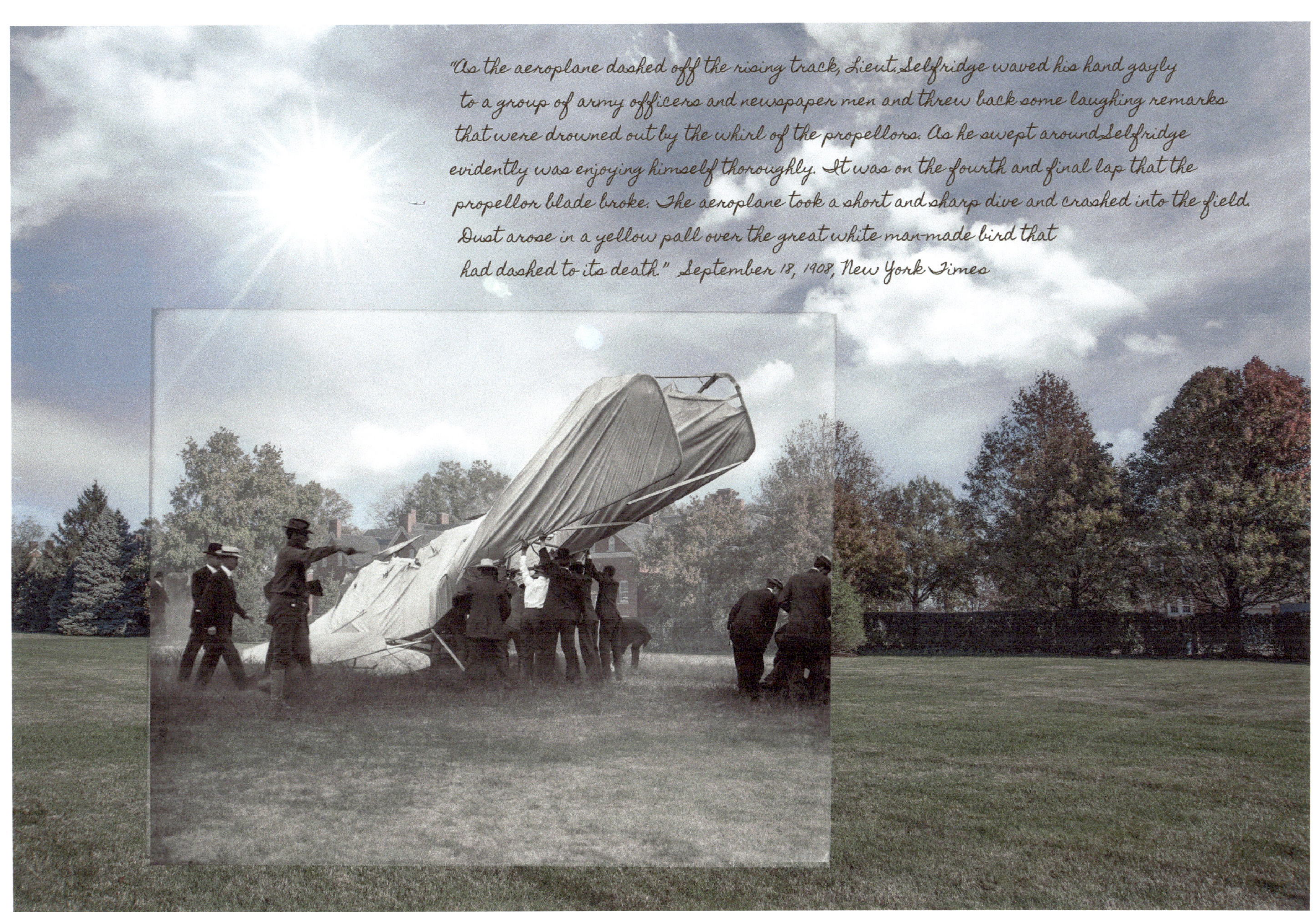

"As the aeroplane dashed off the rising track, Lieut. Selfridge waved his hand gayly
to a group of army officers and newspaper men and threw back some laughing remarks
that were drowned out by the whirl of the propellors. As he swept around Selfridge
evidently was enjoying himself thoroughly. It was on the fourth and final lap that the
propellor blade broke. The aeroplane took a short and sharp dive and crashed into the field.
Dust arose in a yellow pall over the great white man-made bird that
had dashed to its death." September 18, 1908, New York Times

Horse Drawn Carriages

Europe was still eager to see Wilbur fly, so in January 1909, he relocated to Pau,
a city in southwest France. He sent a letter to persuade his brother, Orville, and sister,
Katharine, to cross the Atlantic and join him in France. Wilbur did not realize that
Katharine had already decided they would be joining him as their letters crossed in
the mail. In Pau, not only was the winter weather more hospitable, but it was a resort
destination for the European wealthy who were also eager to see Wilbur fly. The Wright
family were now celebrities.

In his book *The Wright Brothers,* David McCullough states, "Virtually every day but
Sunday a steady stream of elegant carriages and automobiles headed out to the flying
field..." I was so pleased to find this image from Pau of just that. There is no visible
road in the historic photograph, but I liked the idea of creating one. When I combined
the historic photo with my modern image, I decided to make it look like the past's
carriages were traveling on the modern-day road.

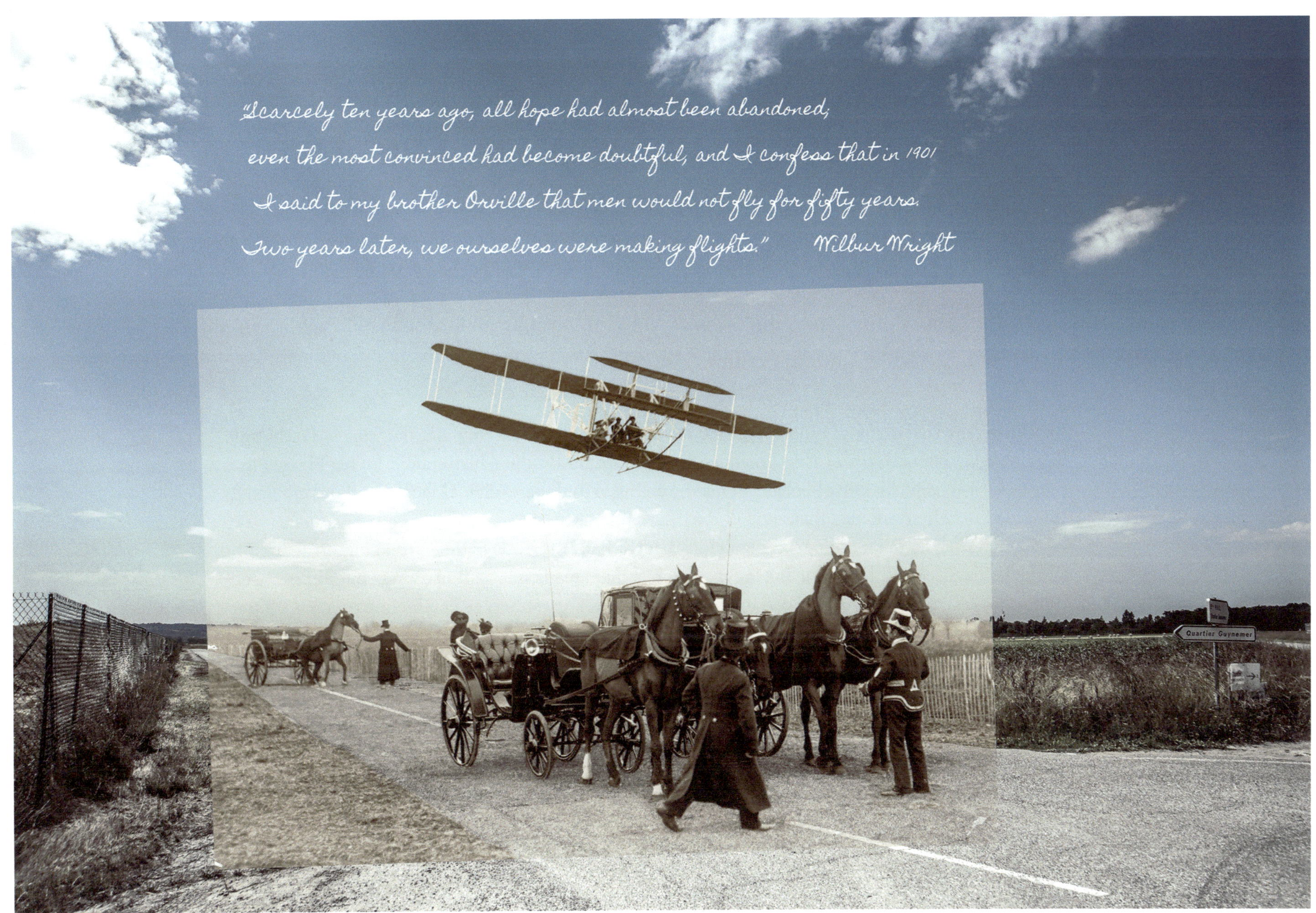

"Scarcely ten years ago, all hope had almost been abandoned;
even the most convinced had become doubtful, and I confess that in 1901
I said to my brother Orville that men would not fly for fifty years.
Two years later, we ourselves were making flights."   Wilbur Wright
Quartier Guynemer

The Infinite Highway Of The Air

"The desire to fly is an idea handed down to us by our ancestors who, in their grueling travels across trackless lands in prehistoric times, looked enviously on the birds soaring freely through space, at full speed, above all obstacles, on the infinite highway of the air."
Wilbur Wright

During my visit to Pau, France, in 2019, I was fortunate to be taken under the wing of Paul Mirat. Paul is the author of *Autrefois Pau L'aviation* and is a Wright Brothers historian in France. I went to France with specific historical images in mind wanting to find their original locations. Paul graciously consented to help me do just that.

We left early in the morning, taking the road out of Pau that leads to the current airport, searching for the fields where Wilbur made his flights. Much of that area is now part of a French military base. While stopped at the side of the road to photograph, French paratroopers conducting training started dotting the sky. I made some images, and we quickly left.

I thought immediately to pair the images I had just taken with this historical image of a farmer staring at the plane flying overhead. I initially loved the historical photo because it compared the old technology of using animals to pull a cart with the newly emerging flight technology. The addition of the paratroopers was better yet. Was the farmer looking at Wilbur's flying machine or the French military plane with people jumping out the back?

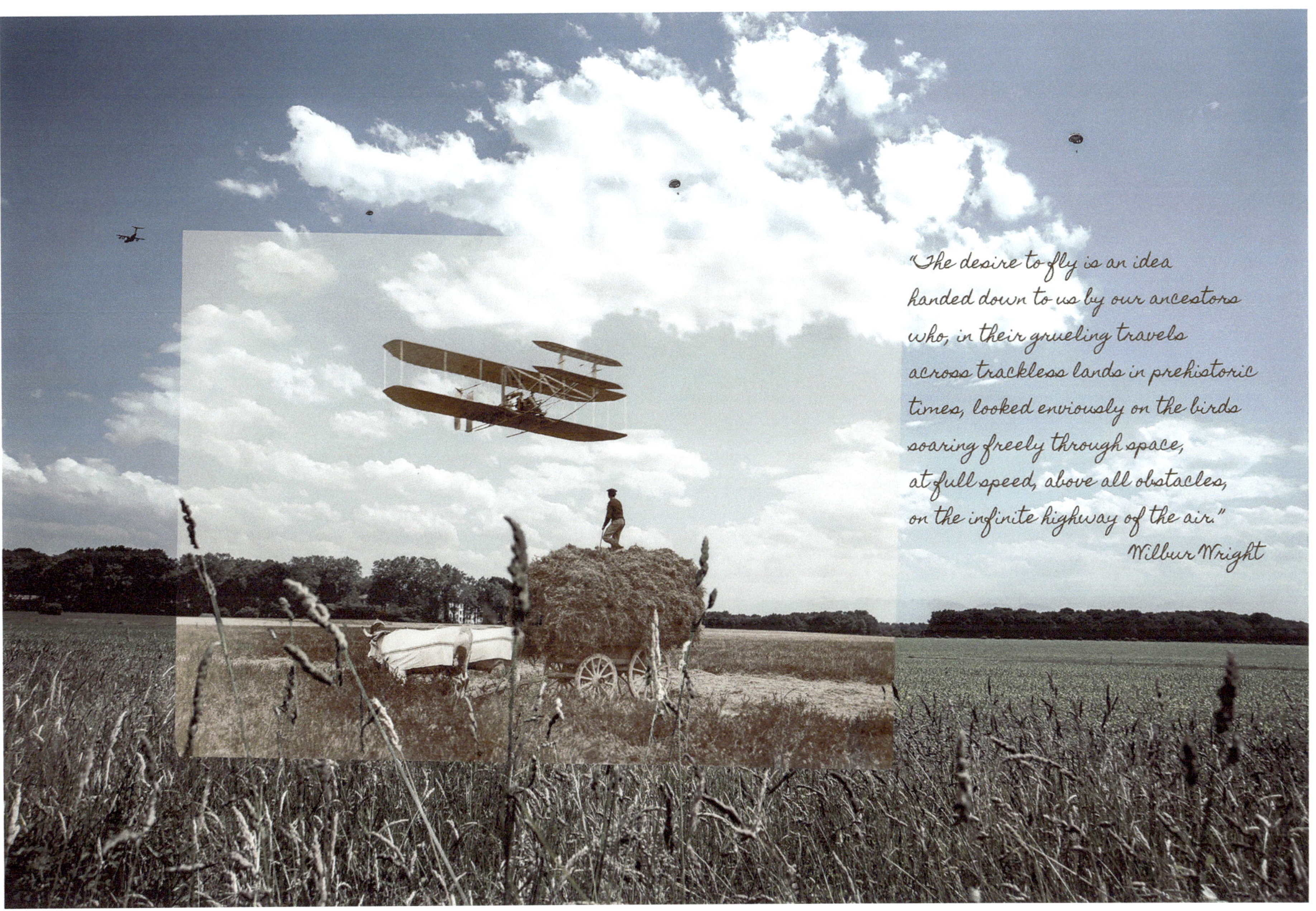
"The desire to fly is an idea
handed down to us by our ancestors
who, in their grueling travels
across trackless lands in prehistoric
times, looked enviously on the birds
soaring freely through space,
at full speed, above all obstacles,
on the infinite highway of the air."
Wilbur Wright

Walking The Boulevard des Pyrénées

"Every time we make a move, the people on the street stop and stare at us.
We have our pictures taken every two minutes."
Katharine Wright

While searching the photo archive at Wright State University in Dayton, Ohio, I came across an image of the Wrights that I knew I wanted to create a piece of artwork around. The picture is of Katharine Wright, Countess Cordelia de Lambert, Orville Wright, and Wilbur Wright strolling along the Boulevard des Pyrénées in Pau, France. By this time, the Wrights had become celebrities and Katharine was enjoying the notoriety.

The Boulevard des Pyrénées is a mile-long walkway that faces the Pyrenees Mountains. My first sight of the Boulevard was as I exited the Pau Funicular that transports people up the hill from the train station to the old city center. I was very excited to see that the Boulevard had not changed in the 110 years since the historic photograph was taken. The cobblestone and the railing were all original.

I stayed in the historic part of downtown Pau and the Boulevard was very close to my hotel. I walked up and down the Boulevard at different times of the day to capture photographs to use. I took more than 400 pictures of the Boulevard, knowing that it would be difficult to match my image with the historical one precisely.

I did eventually find an image that was an excellent match to the original composition. Unfortunately, my first attempt at creating an image was boring. It had no additional people in it. I went back through my photographs and selected other photos that included people and even a bird that flew through the camera frame. With these additions, I was able to make the Boulevard come alive.

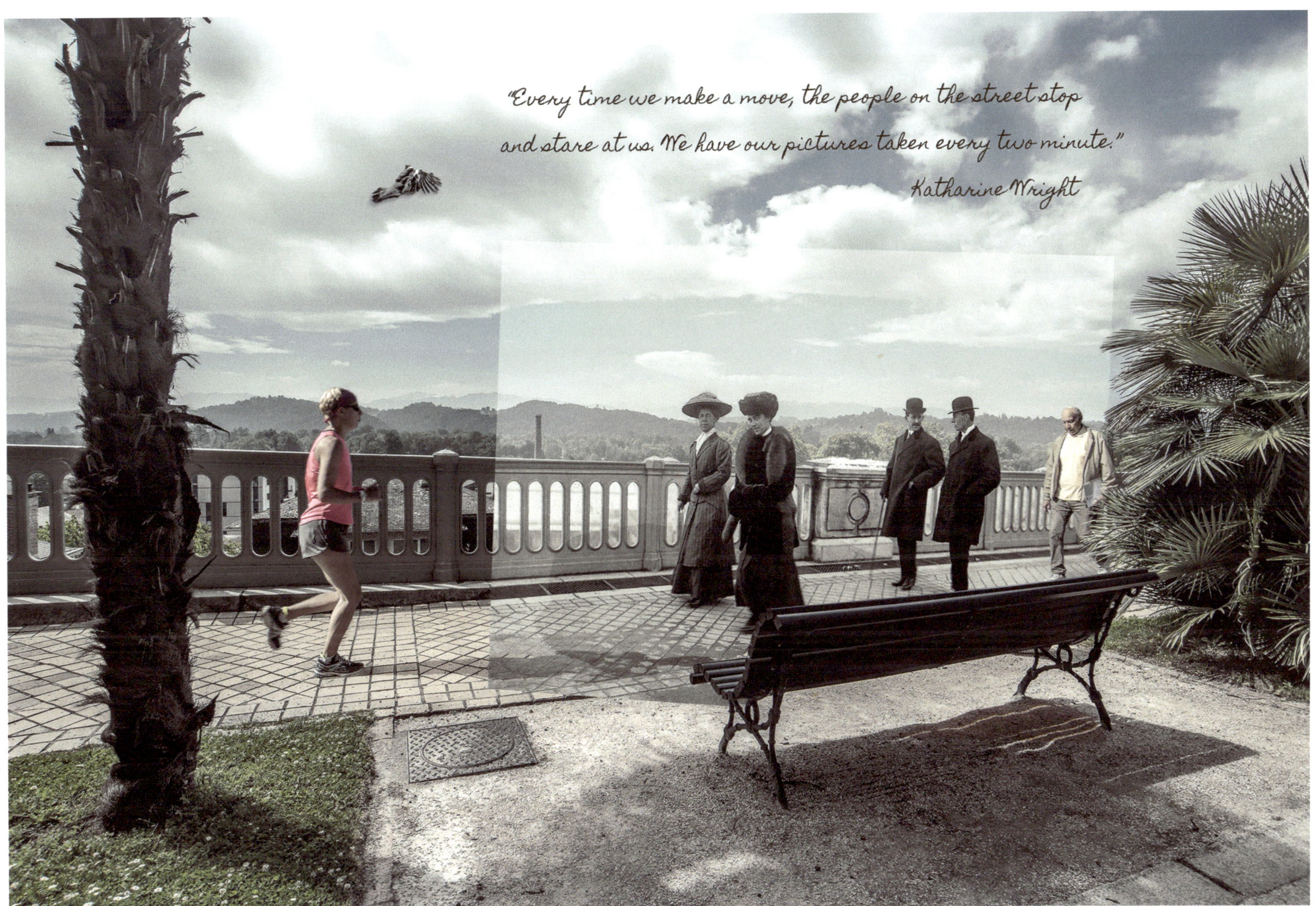

"Every time we make a move, the people on the street stop and stare at us. We have our pictures taken every two minute."
Katharine Wright

Greenfield Village

The photograph of Wilbur and Orville was taken in June of 1909. The brothers are seated on the back porch of their home at 7 Hawthorn Street in Dayton, Ohio. Wilbur spent the previous year in Europe, thrilling the world with his flight demonstrations. Orville made successful test flights for the government in Washington, D.C. The two men were now world-famous.

In 1937, Henry Ford purchased the Wright Brothers' house at 7 Hawthorn Street and moved it to Greenfield Village in Dearborn, Michigan. I knew I wanted to make some images of the original house so, in the fall of 2019, my wife, Maria, and I made the trip to Dearborn. It was amazing to walk around to the rear of the house and see the porch looking precisely as if no time had passed since the original image was made. Equally important to me was the backyard shed, which was also there. The Wright Brothers used the shed as their darkroom. They processed all their film and made prints of their flight experiments, including the famous photograph of the first powered flight at Kitty Hawk on December 17, 1903, in that building.

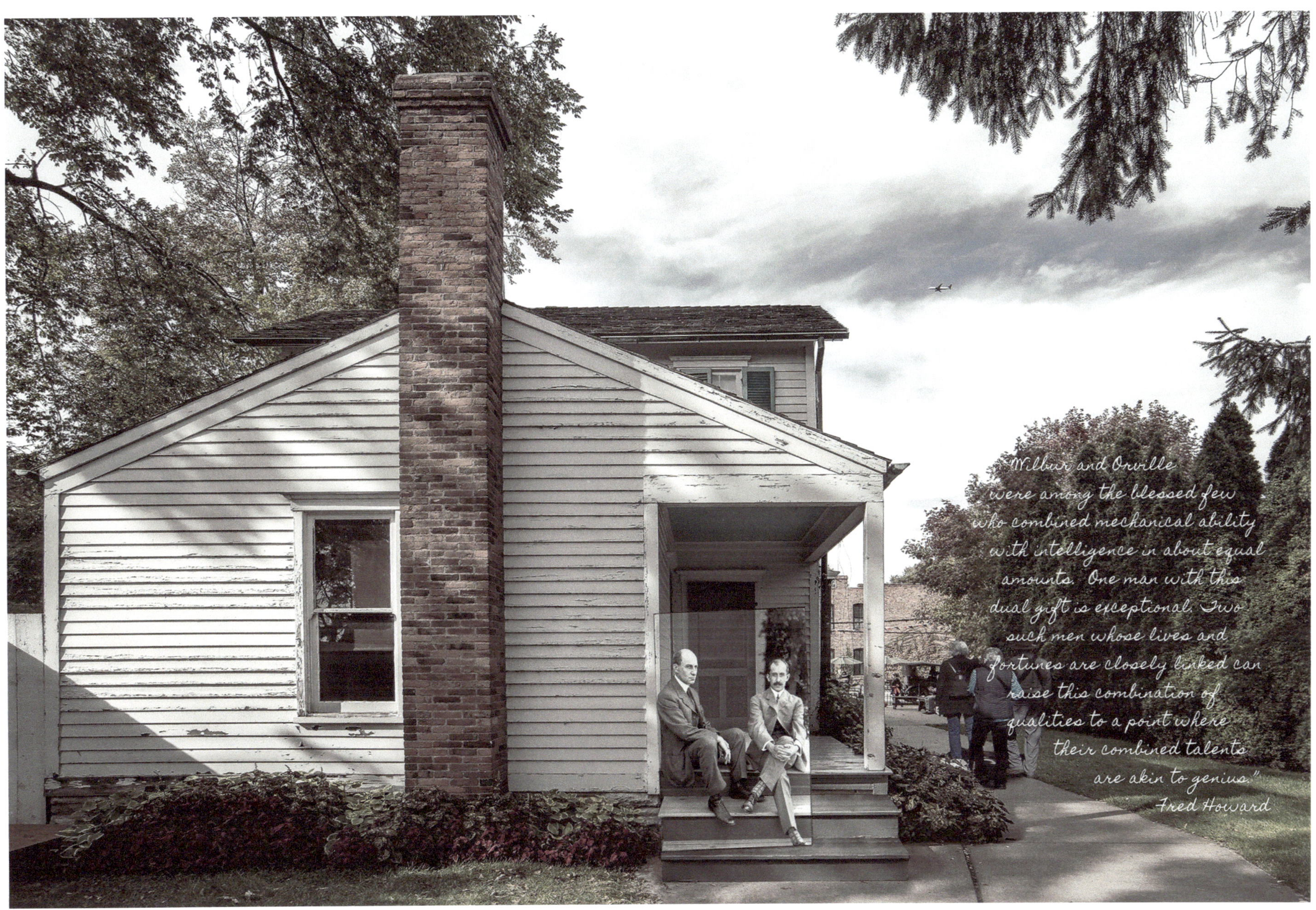
Wilbur and Orville
were among the blessed few
who combined mechanical ability
with intelligence in about equal
amounts. One man with this
dual gift is exceptional. Two
such men whose lives and
fortunes are closely linked can
raise this combination of
qualities to a point where
their combined talents
are akin to genius."
Fred Howard

The Hudson-Fulton Celebration

In the fall of 1909, New York and New Jersey had a commemoration of the 300th anniversary of Henry Hudson's discovery of the Hudson River and Robert Fulton's 100th anniversary of the paddle steamer. It was called the Hudson-Fulton Celebration. Wilbur Wright and Glenn Curtiss, the two most celebrated pilots of the day, gave flight demonstrations on the Hudson River as part of the celebration. The conditions were very windy, and Glenn Curtiss only made a couple of brief flights. However, Wilbur made numerous flights, and on September 29, 1909, he flew around the Statue of Liberty. On October 4th, he made a 33-minute flight up the Hudson River to Grant's tomb, where hundreds of thousands of New Yorkers saw him fly.

April of 2020 was when I originally planned to travel to New York City and create photographs for this series. Then the pandemic hit, and travel halted nationwide. On September 15, 2020, travel between New York, New Jersey, and Ohio opened again, and I made a quick trip to create photographs. On October 20th, I drove from Ohio to Newark, New Jersey. The morning I went into Manhattan, there was a dense fog over the city that didn't lift until about 3 p.m.

I took the New York Waterway ferry transport boat from New Jersey to Manhattan. I captured this image from the back of the ship as it traveled across the Hudson River. On the left side of my photo is New Jersey, and the right is the Manhattan skyline. In the historic photograph, Wilbur is flying up the Hudson River past the military battleships and over the yacht *Viking*, owned by New York banker George Baker, Jr. I liked seeing the steam coming from the ships and thought this quote from a newspaper interview worked well within my image.

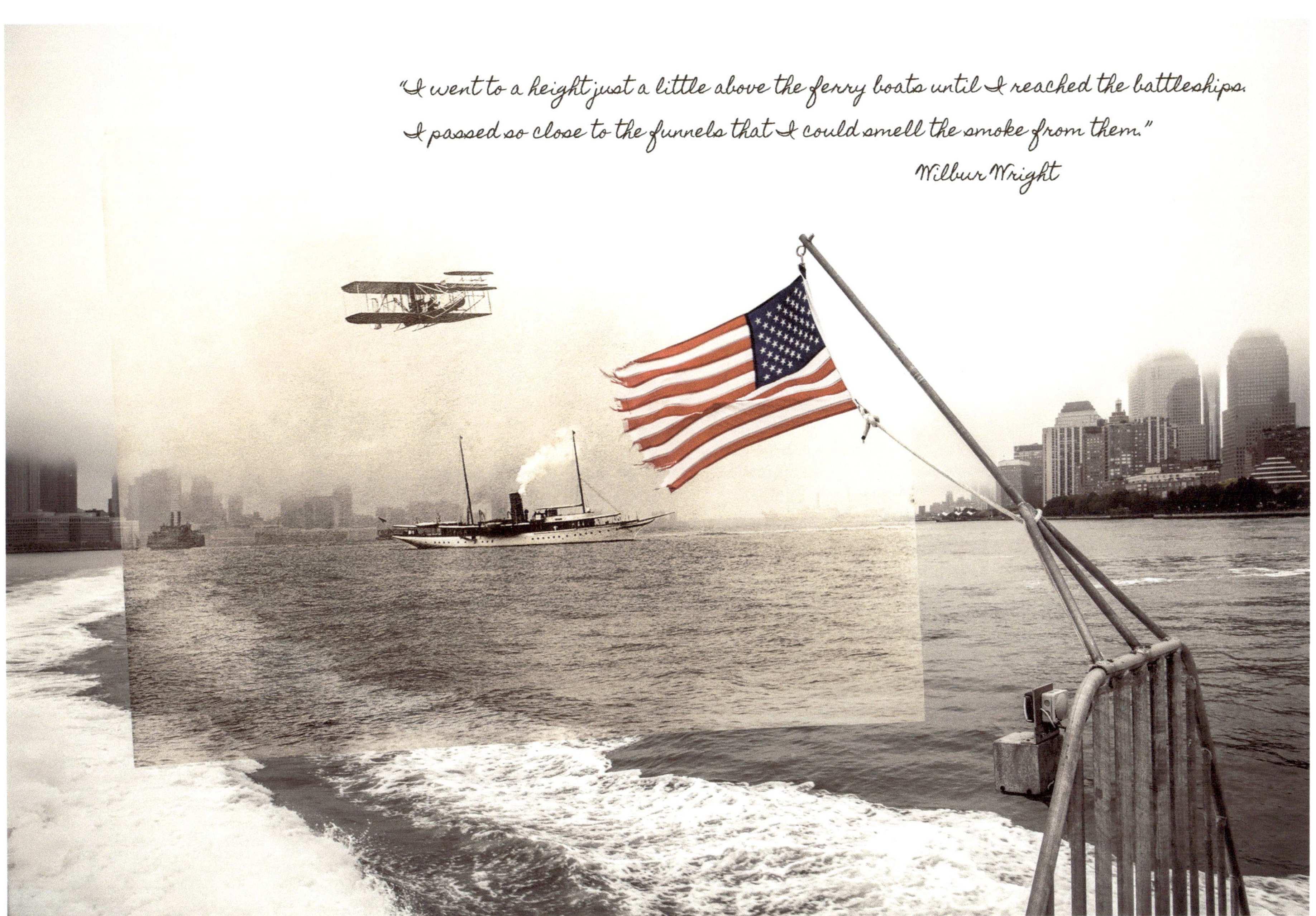

"I went to a height just a little above the ferry boats until I reached the battleships.
I passed so close to the funnels that I could smell the smoke from them."
Wilbur Wright

Lady Liberty

The Staten Island ferry goes between Manhattan and Staten Island and travels right past the Statue of Liberty. I wasn't happy with the photographs I created on my Staten Island ferry ride because the Statue of Liberty was too far in the distance. I then decided to take the Statue of Liberty boat from Battery Park out to Liberty Island. I captured this photo from the upper deck as the ferry was arriving at Liberty Island. Even before making the trip to New York, I envisioned a photograph just like this one with people looking out at the Statue of Liberty. I am thrilled with the way it turned out. I love how the four young ladies are posing for a selfie, and the gentleman is taking pictures as if he can see Wilbur flying around the Statue of Liberty.

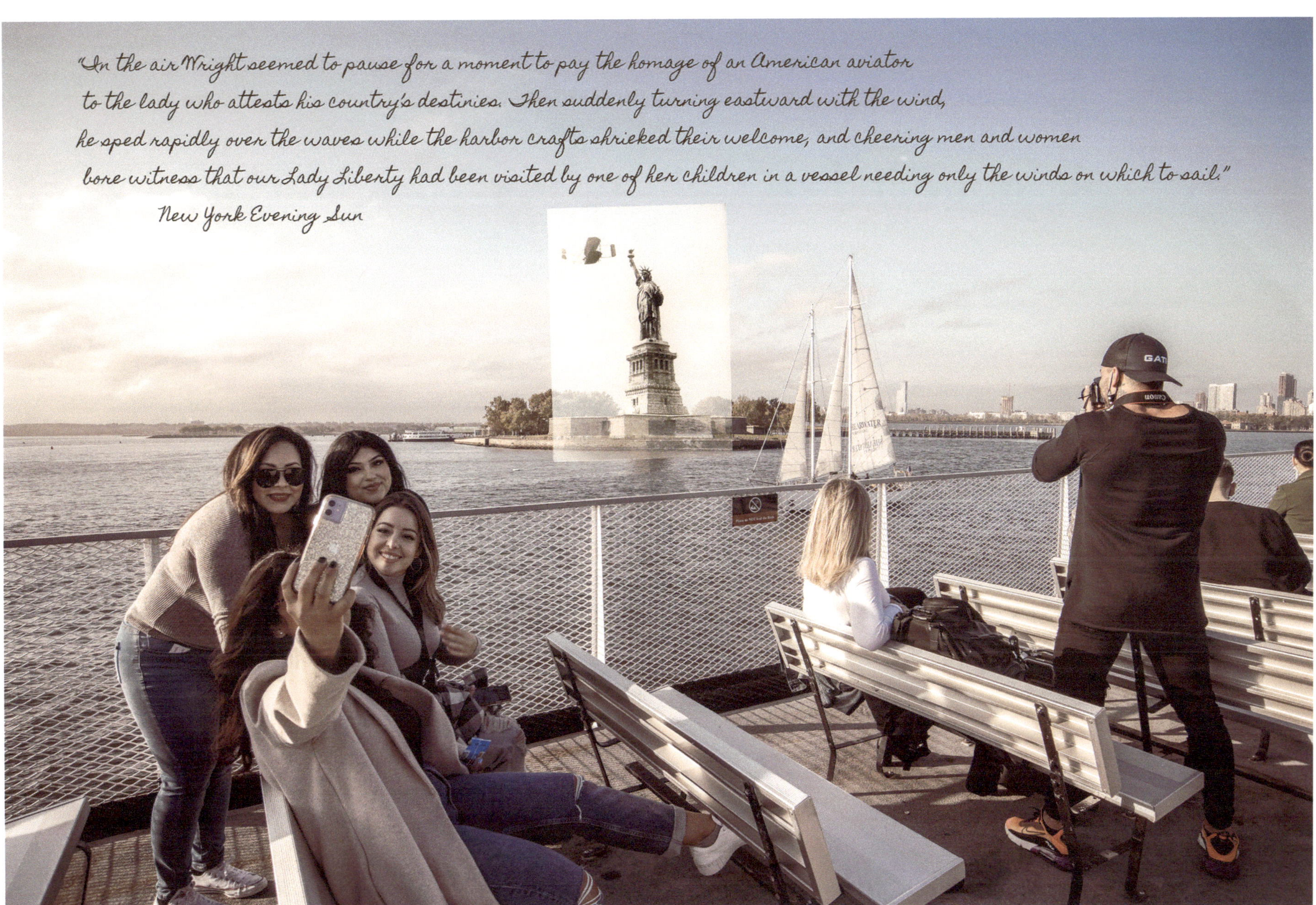
"In the air Wright seemed to pause for a moment to pay the homage of an American aviator
to the lady who attests his country's destinies. Then suddenly turning eastward with the wind,
he sped rapidly over the waves while the harbor crafts shrieked their welcome, and cheering men and women
bore witness that our Lady Liberty had been visited by one of her children in a vessel needing only the winds on which to sail."
New York Evening Sun

A New Kind Of Gull In New York Harbor

The title for this image comes from the cover of *Harper's Weekly* on October 9, 1909. *Harper's Weekly* had a photograph of Wilbur flying around the Statue of Liberty on the cover with the title "A New Kind Of Gull Flying In New York Harbor." This photograph was the last image I created for this series.

I am a photographer who believes in serendipity: "The occurrence of events by chance in a happy or beneficial way." I had been in lower Manhattan since early morning. The last thing I did that day was take The Statue of Liberty-Ellis Island tour. I was on Ellis Island waiting for the last ferry of the day off the island. I was initially planning to go back to Manhattan. There were two boats at the Ellis Island dock ready for departure: one to Manhattan and one to New Jersey. I was tired, and it was getting late in the day, so I took the boat back to New Jersey. I created this photograph on the upper deck at the back of the ferry. The sun was going down and cast a beautiful light on the skyline of Manhattan. If I had taken the ferry back to Manhattan, I would have missed the beautiful glow of the setting sun, and it would have been dark when I crossed the Hudson River. When I looked through my photographs, deciding which to use, I found this image with a seagull flying through my frame, and the title from *Harper's Weekly* immediately came to mind.

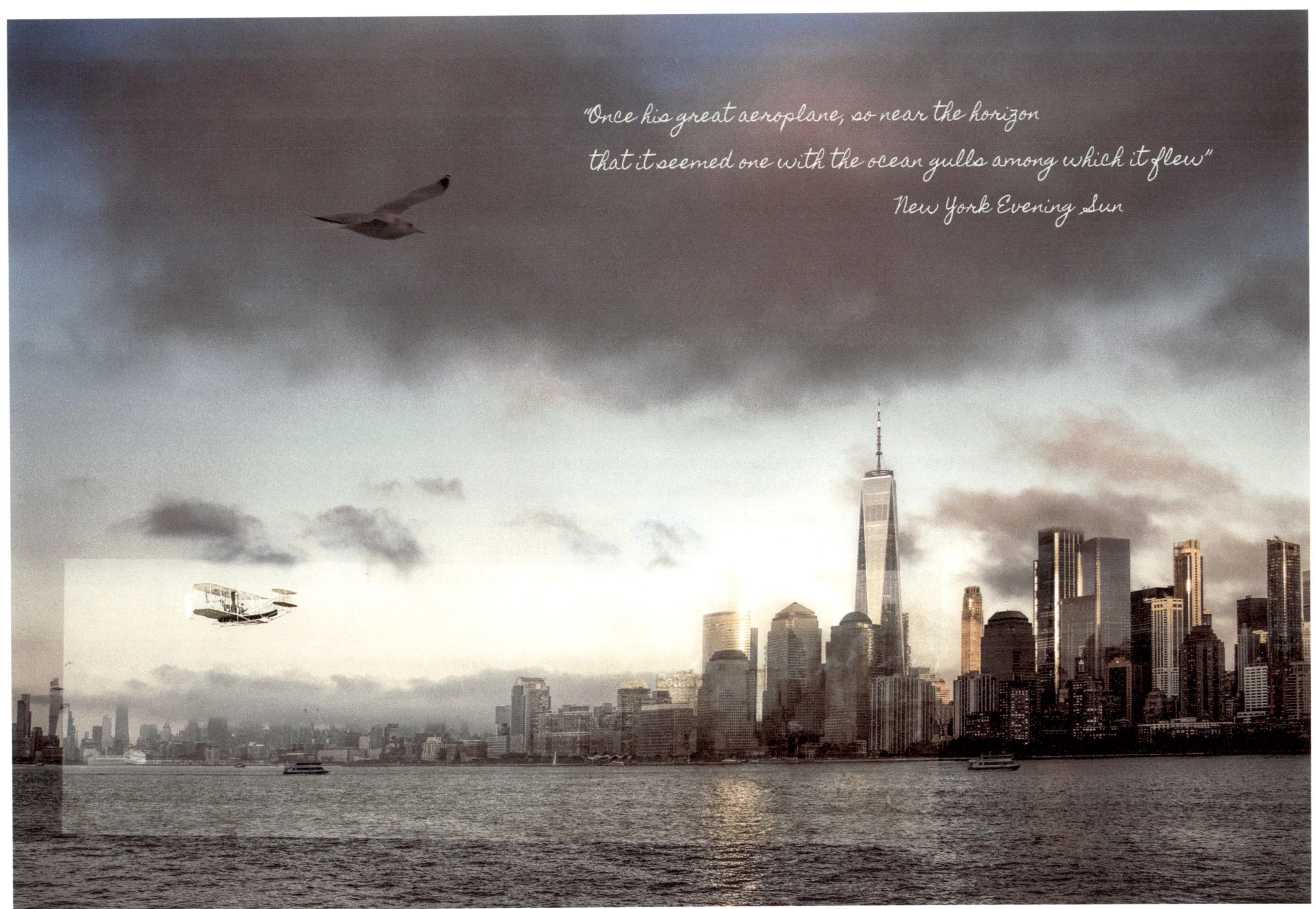
"Once his great aeroplane, so near the horizon
that it seemed one with the ocean gulls among which it flew"
New York Evening Sun

Bishop's First Flight

"Higher, Orville, Higher!"
Bishop Milton Wright

In many cases, my starting point when creating images for this series is with the historic photograph. For *Bishop's First Flight,* my inspiration was the very moving passage in David McCullough's book on the Wright Brothers that described the day on May 25, 1910, when friends, neighbors, and family were invited to Huffman Prairie for a flying demonstration.

McCullough points out that Wilbur and Orville had never flown together, ensuring that the other could continue their work in case of a tragedy. But on this day, both Orville and Wilbur flew together over Huffman Prairie. It was also the day their father, Bishop Wright, at age 85, flew for the first time and was reported to have said, "Higher, Orville, higher!" while in the air. They soared over Huffman Prairie at around 350 feet for about six minutes.

Huffman Prairie Flying Field National Historical Park is located next to Wright Patterson Air Force Base. As I took photos that day, a *C-17 Globemaster III* air transport took off and flew through my camera's frame. The *C-17* was at the same altitude in my image as Orville and Milton Wright were in 1910. The quote "Higher, Orville, higher" for me is a metaphor for where aviation was soon to be.

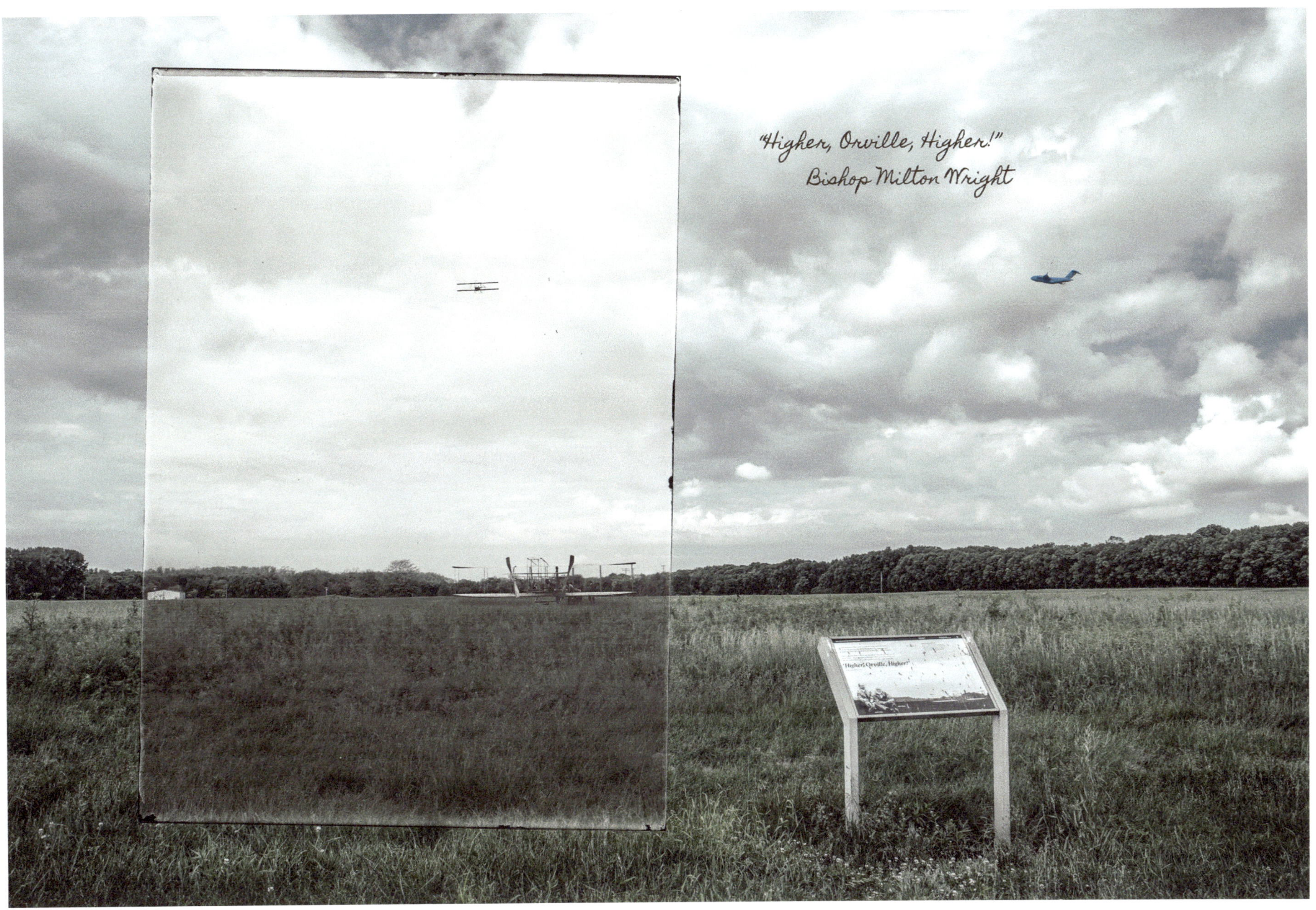

"Higher, Orville, Higher!"
Bishop Milton Wright

Test Flight On The Miami River

"Isn't it astonishing that all these secrets have been preserved for so many years just so we could discover them!"
Orville Wright

For many years I rode my bike from West Carrollton, Ohio, where I lived, to Carillon Park in downtown Dayton along the Great Miami River bike path. Often, I would stop for a drink of water at a boat put-in on the river bike path, rest for a while, and watch the river flow past. Years later, while researching historical images for my project, I came across this photograph. I recognized it at once as the very same place I would stop while riding my bike.

The photograph is attributed to Preston Mayfield of the Dayton Daily Newspaper and was taken on May 1, 1913. Mayfield may have been the first person to take an aerial photograph while flying with Orville Wright. Two men are sitting in the boat on the river and observing the *Wright Model CH Flyer* fitted with twin, multi-step pontoons fly overhead. I believe Orville is watching the test flight from the boat. This part of the Miami River flows east-west and then makes a sharp turn going north-south. It allowed test flights to be made into the wind going in either direction.

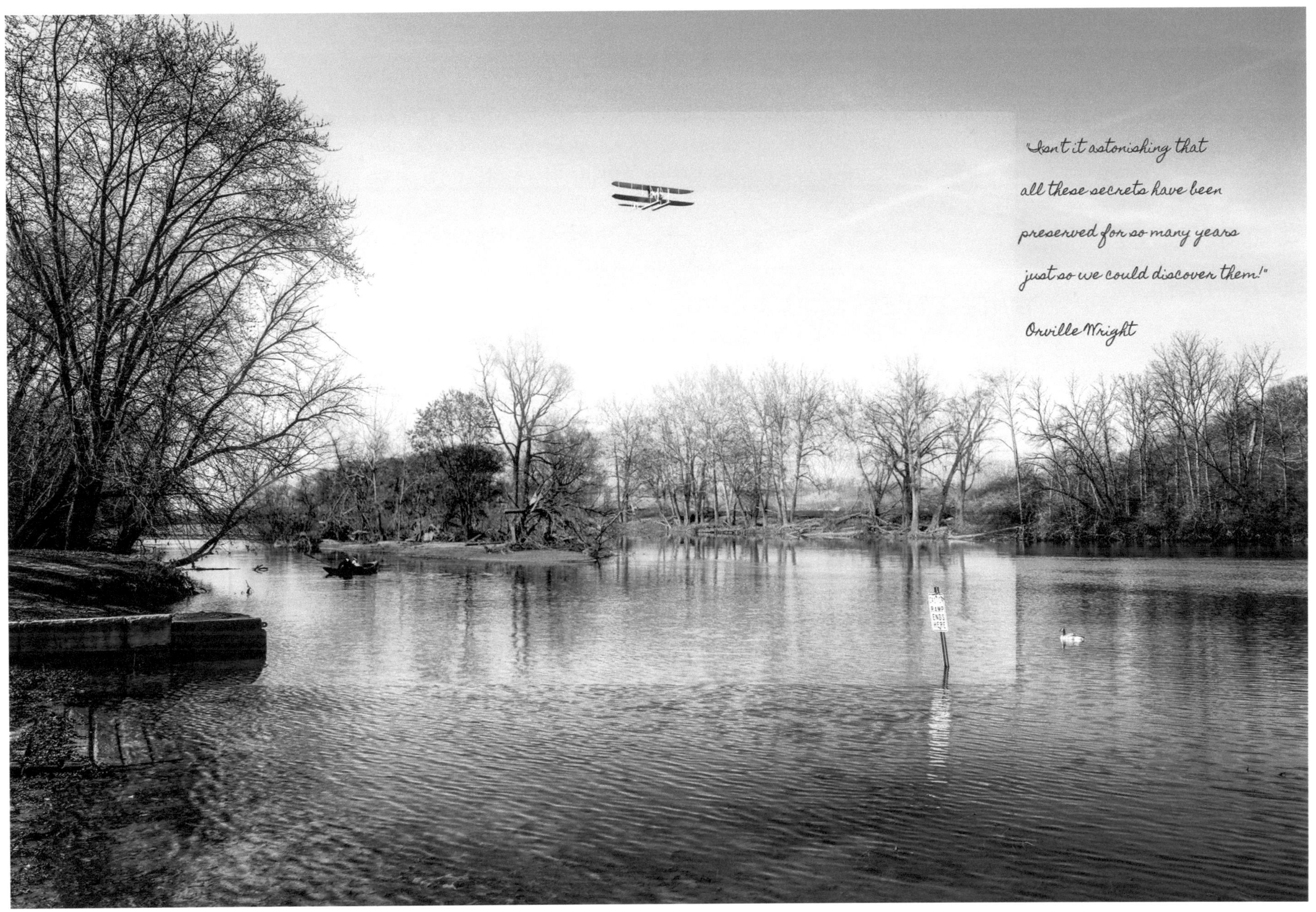
Isn't it astonishing that
all these secrets have been
preserved for so many years
just so we could discover them!"

Orville Wright

Family

"But it isn't true to say we had no special advantage, the greatest thing in our favor was growing up in a family where there was always much encouragement to intellectual curiosity."
Orville Wright

In February 1912, the Wrights purchased a 17-acre lot in Oakwood, Ohio, a Dayton suburb. They hired architects Schenck & Williams to design the new home they planned to build on the site. Unfortunately, before construction on the house could begin, Wilbur Wright contracted typhoid fever and died on May 30, 1912.

After Wilbur's death, Orville and Katharine continued with the new home plans, and in August of that year, construction began. It was completed in the spring of 1914, and on April 28, Bishop Wright and Katharine moved in. Orville, on a business trip, would follow two days later. The new home became known as Hawthorn Hill for the many hawthorn trees covering the hillside and a reminder of their boyhood home on Hawthorn Street in Dayton.

The historic photograph was taken in 1915. It shows Bishop Wright in the center with Orville on his right and Katharine on his left. In the photo's far left is Horace Wright, nephew of Orville, and Katharine and Lorin Wright's son. The other gentlemen are Earl N. Findley, an early aviation reporter, John R. McMahon, who wrote The Wright Brothers: Fathers of Flight, and Pliny Williamson, the Wright Brothers' lawyer.

I chose the title *Family* for this image for two reasons. First, Bishop Wright's importance on family ties continued with Orville and Katharine, who remained close to their nieces and nephews, and their children. Hawthorn Hill became a place for the extended family to gather and spend the day with Aunt Katharine and Uncle Orv. Secondly, the people in the historic photo are a combination of family and friends. Yet, the image feels just like a family portrait with everyone sitting on the lawn posing with the house in the background. I wanted to play with time in this image. Is the tree in the frame old or new? Did the driveway exist in the historic photograph?

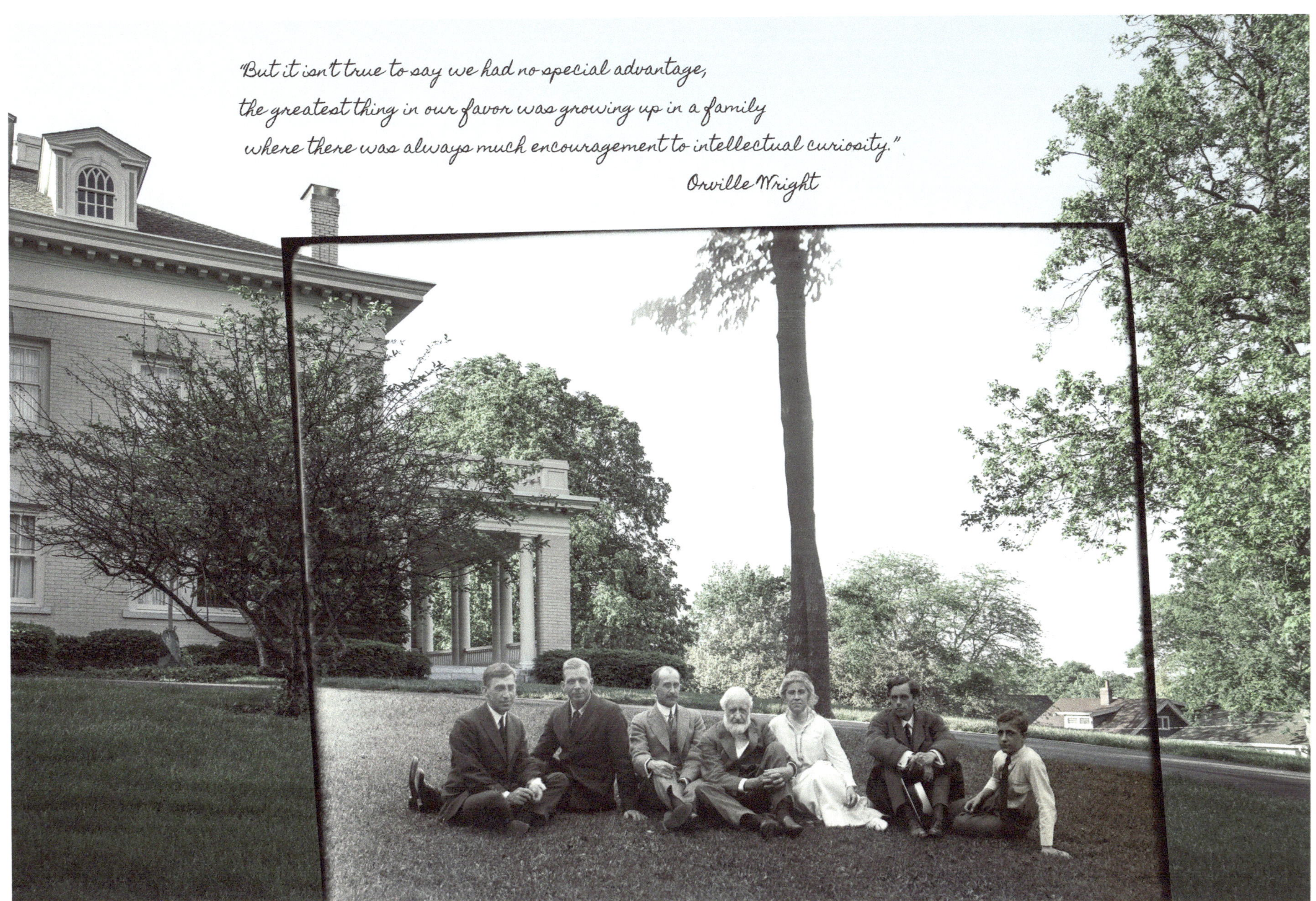

"But it isn't true to say we had no special advantage,
the greatest thing in our favor was growing up in a family
where there was always much encouragement to intellectual curiosity."
Orville Wright

Scipio

"When Orville Wright dies in 1948, they only found one photograph in his wallet. It wasn't of the first flight, or of his sister or parents or of he and his brother. It was a photograph of his dog Scipio who had died 25 years earlier."
Stephen Wright

Milton Wright's diary entry of March 10, 1917, states, "Scipio came. He weighs 16 pounds. He is a St. Bernard dog. He is a good-looking puppy." Orville purchased the dog from a breeder in New Jersey for $75, and he soon became a member of the family. Katharine named him Scipio after the famous Roman general who had defeated Hannibal and thwarted an invasion of Rome. Orville loved his dog and took many photographs of him. What I especially liked about this historic photograph is the expression on Scipio's face. It is the face of a dog who loves the photographer, his owner. I also loved the black air bubble marks and uneven edges of the negative that are evident in the historic image. For me, they are photographic "beauty marks" and are evidence of film processed by hand in a home darkroom.

I created many photographs on the porch at Hawthorn Hill, knowing one would work with this historical image, but it took me several months of research to find the perfect quote. Looking online, I found a podcast from WYSO Weekend, our local NPR station, from April 17, 2016. I heard the story about Scipio and Orville and knew I found the quote to use. Orville Wright died January 30, 1948 and only one photograph was found in his wallet. It was of Scipio, his beloved St. Bernard, who had preceded him in death by nearly 25 years.

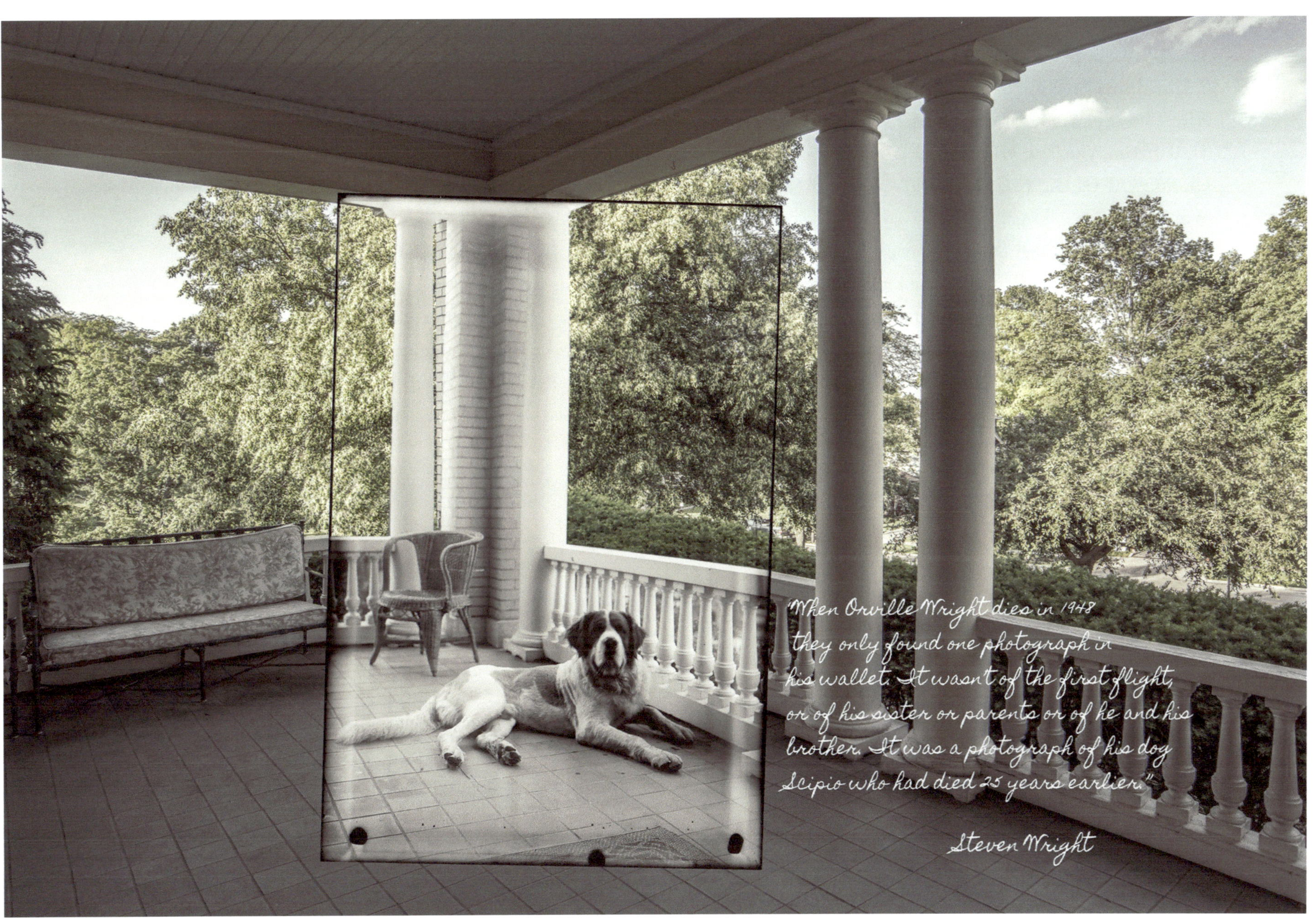
When Orville Wright dies in 1948
they only found one photograph in
his wallet. It wasn't of the first flight,
or of his sister or parents or of he and his
brother. It was a photograph of his dog
Scipio who had died 25 years earlier."

Steven Wright

Locations of Current Day Photographs

The Bike Shop – Wright Brother's Bicycle Shop, 22 South Williams Street, Dayton, Ohio

7 Hawthorn Street – In front of 7 Hawthorn Street, Dayton, Ohio

Kite Flying – Wright Brothers National Memorial, 1401 National Park Drive, Manleo, North Carolina

View From The Top Of Big Kill Devil Hill - Wright Brothers National Memorial, 1401 National Park Drive, Manleo, North Carolina

Gliding Off Of Big Kill Devil Hill - Wright Brothers National Memorial, 1401 National Park Drive, Manleo, North Carolina

Waiting To Fly - Wright Brothers National Memorial, 1401 National Park Drive, Manleo, North Carolina

First Flight - Wright Brothers National Memorial, 1401 National Park Drive, Manleo, North Carolina

Brothers – Huffman Prairie Flying Field National Historic Park, Pylon Road, Wright-Patterson Air Force Base, Ohio

Flight #85 - Huffman Prairie Flying Field National Historic Park, Pylon Road, Wright-Patterson Air Force Base, Ohio

Flight #41 - Huffman Prairie Flying Field National Historic Park, Pylon Road, Wright-Patterson Air Force Base, Ohio

Traveling With The Wind - Huffman Prairie Flying Field National Historic Park, Pylon Road, Wright-Patterson Air Force Base, Ohio

Amos Root - Huffman Prairie Flying Field National Historic Park, Pylon Road, Wright-Patterson Air Force Base, Ohio

Le Mans Racetrack - Place Luigi Chinetti, Le Mans, France

King Alfonso XIII Of Spain – Le Mans Racetrack, Place Luigi Chinetti, Le Mans, France

Speed Test At Fort Myer – Fort Myer Army Base, Arlington, Virginia

The Crash – Fort Myer Army Base, Arlington, Virginia

Infinite Highway Of The Air – D816, Pau, France

Horse Drawn Carriages - D816, Pau, France

Walking The Boulevard des Pyrénées - Pau, France

Greenfield Village – Dearborn, Michigan

The Hudson-Fulton Celebration – New York City, New York

Lady Liberty – New York City, New York

A New Kind Of Gull In New York Harbor – New York City, New York

Bishop's First Flight - Huffman Prairie Flying Field National Historic Park, Pylon Road, Wright-Patterson Air Force Base, Ohio

Test Flight On The Miami River – Great Miami River Bike Trail, Dayton, Ohio

Family – Hawthorn Hill, 901 Harman Avenue, Oakwood, Ohio

Scipio - Hawthorn Hill, 901 Harman Avenue, Oakwood, Ohio

Historic Photographs

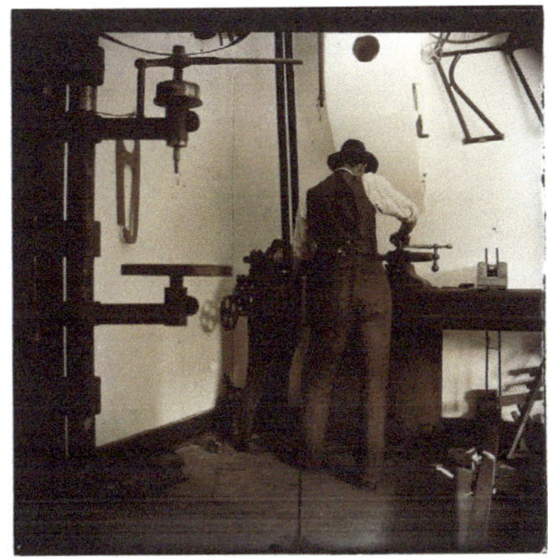

The Bike Shop
Library of Congress, Wright, Wilbur, and Orville Wright, photographer. Wilbur Wright working in the bicycle shop. Ohio Dayton, 1897. Photograph. https://www.loc.gov/item/2001696435/.

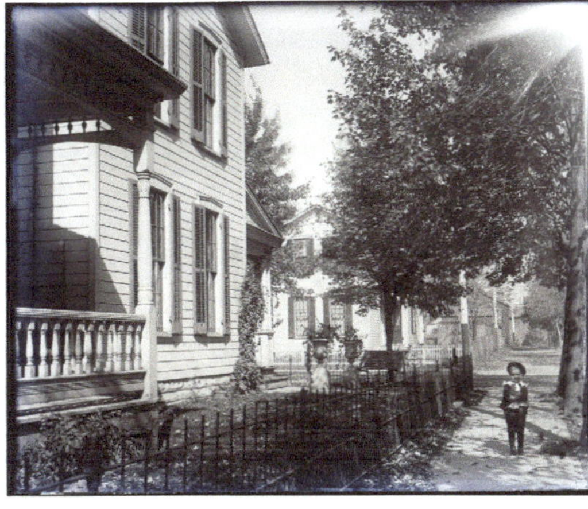

7 Hawthorn Street
Library of Congress, Wright, Wilbur, and Orvill Wright, photographer. Daniel Henderson, neighbor, in front of Wright home at 7 Hawthorn Street, Dayton, Ohio. Ohio Dayton, None. [Between 1897 and 1901] Photograph. https://www.loc.gov/item/2001696283/.

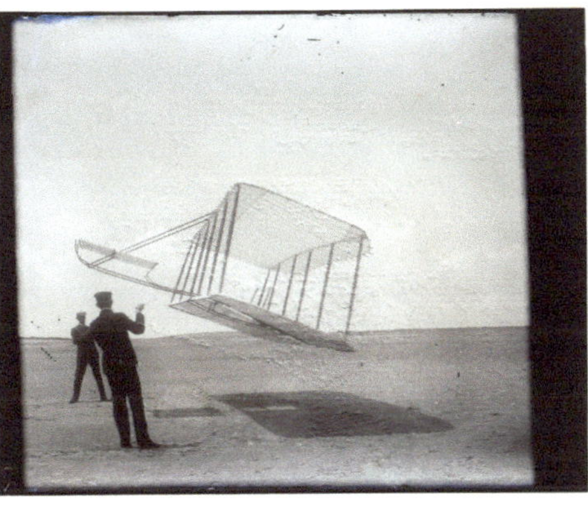

Kite Flying
Library of Congress, Wright, Wilbur, and Orville Wright, photographer. Side view of glider flying as a kite near the ground, Wilbur at left and Orville at right, glider turned forward to right and tipped downward. Kitty Hawk

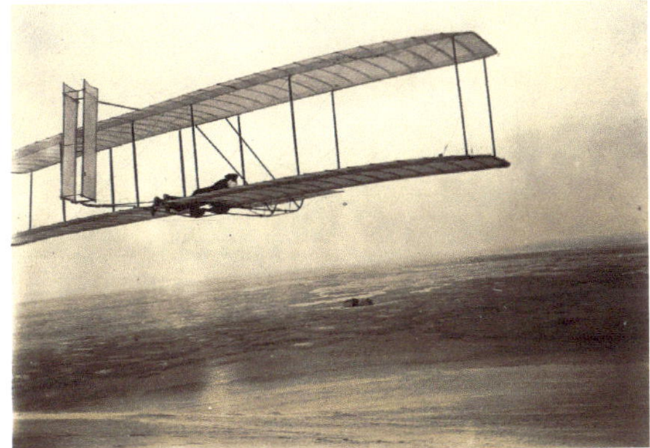

View From The Top Of Big Kill Devil Hill
Wright State University, Item Identifier Number: ms1_15_6_10, Creation Date 10-1-1902, Collection MS-1: Wright Brothers Collection, Description: Wilbur Wright piloting the Wright 1902 glider at Big Kill Devil Hill. This photograph is attributed to Lorin Wright., Publisher Repository: Special Collections and Archives; Wright State University Libraries, Digital Services Department; Wright State University Libraries

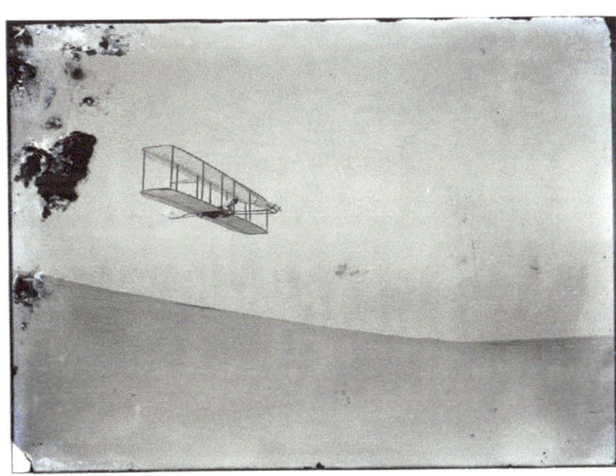

Gliding Off Big Kill Devil Hill
Library of Congress, Wright, Wilbur, and Orville Wright, photographer. Wilbur gliding down steep slope of Big Kill Devil Hill; Kitty Hawk, North Carolina. Kitty Hawk North Carolina, 1902. Photograph. https://www.loc.gov/item/2001696265/.

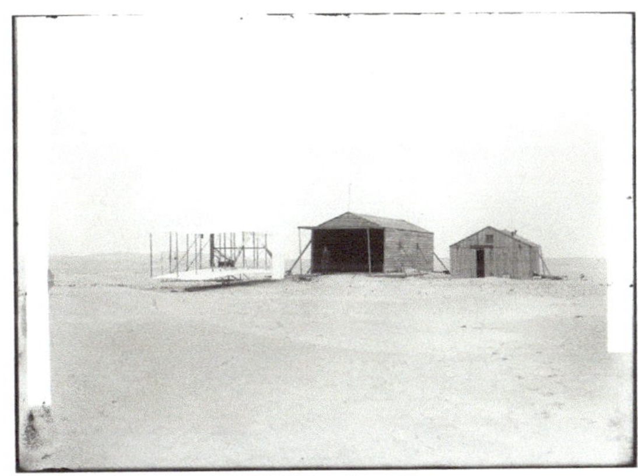

Waiting To Fly
Library of Congress, Wright, Wilbur, and Orville Wright, photographer. machine and large camp building where it was housed, and smaller building used as a workshop and living quarters at Kill Devil Hills. Kitty Hawk North Carolina, 1903. Photograph. https://www.loc.gov/item/2001696490/.

Historic Photographs

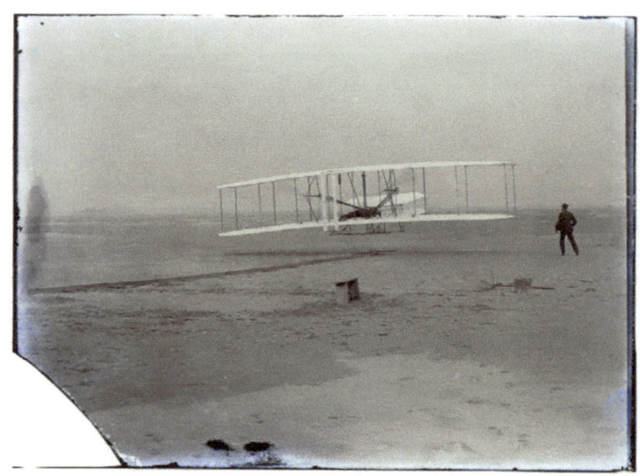

First Flight
Library of Congress, Wright, Wilbur, Orville Wright, and John T Daniels, photographer. First flight, 120 feet in 12 seconds, 10:35 a.m.; Kitty Hawk, North Carolina. Kitty Hawk North Carolina, 1903. Photograph. https://www.loc.gov/item/00652085/.

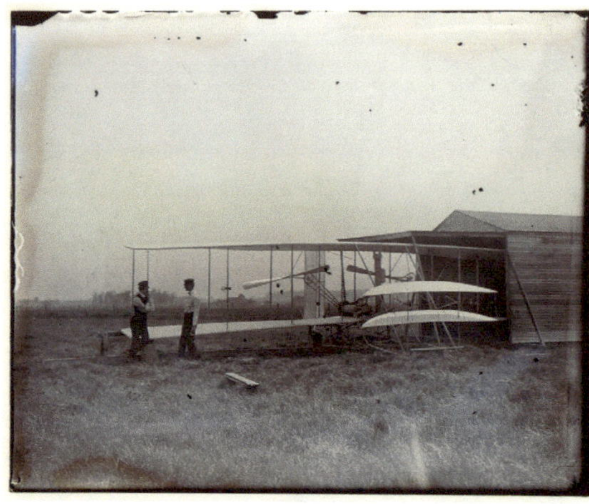

Brothers
Library of Congress, Wright, Wilbur, and Orville Wright, photographer. Wilbur and Orville Wright with their second powered machine; Huffman Prairie, Dayton, Ohio,1904. [May] Photograph. https://www.loc.gov/item/2001696552/.

Amos Root
Library of Congress, Wright, Wilbur, and Orville Wright, photographer. Left front view of flight 46, Orville turning to the left, in the last photo-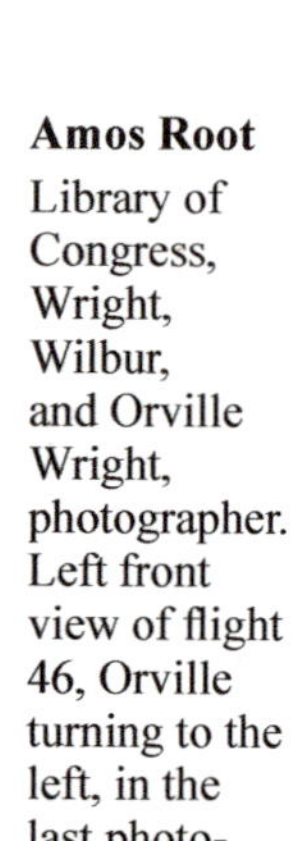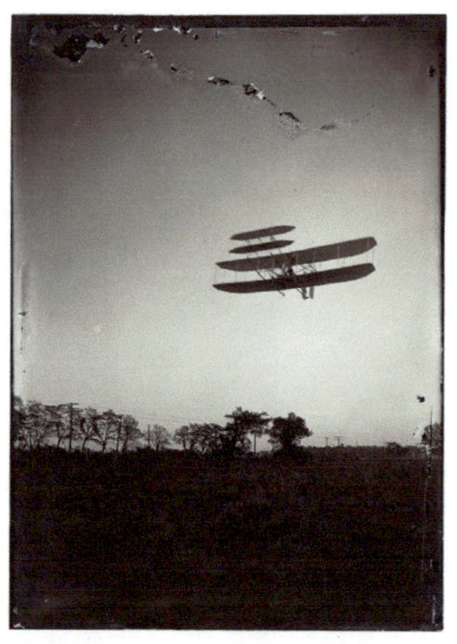graphed flight of ; Huffman Prairie, Dayton, Ohio. Ohio Dayton, 1905. [Oct. 4] Photograph. https://www.loc.gov/item/2001696588/.

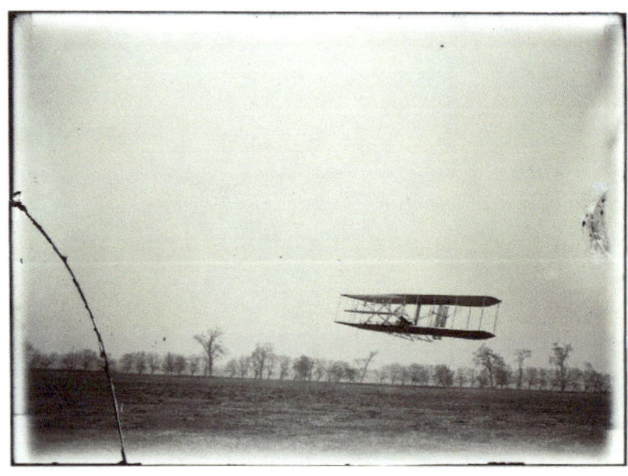

Flight #85
Library of Congress, Wright, Wilbur, and Orville Wright photographer. Flight 85: Orville in flight over treetops, covering a distance of approximately 1,760 feet in 40 1/5 seconds; Huffman Prairie, Dayton, Ohio. Ohio Dayton, 1904. Photograph. https://www.loc.gov/item/2001696497/.

58

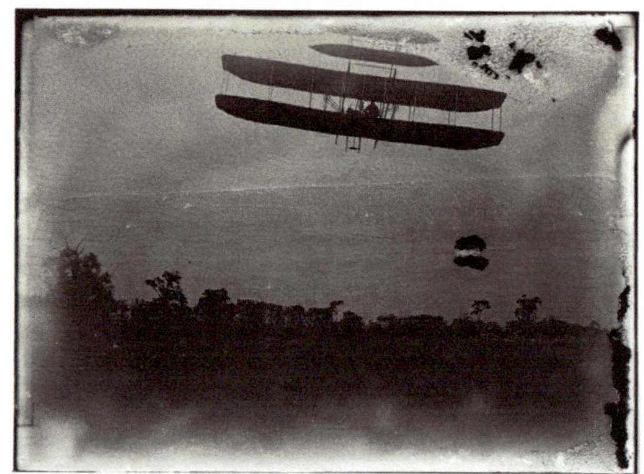

Flight #41
Library of Congress, Wright, Wilbur, and Orville Wright, photographer. Flight 41: Orville flying to the left at a height of about 60 feet; Huffman Prairie, Dayton, Ohio. Ohio Dayton, 1905. [Sept. 29] Photograph. https://www.loc.gov/item/2001696562/.

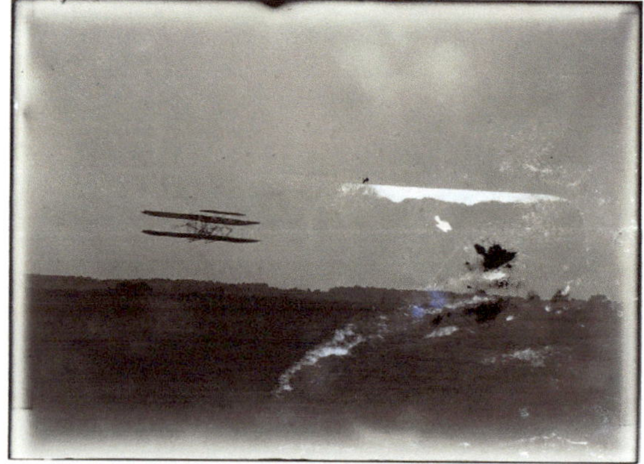

Traveling With The Wind
Library of Congress, Wright, Wilbur, and Orville Wright, photographer. Flight 23: front view of the machine in flight to the right, Orville at the controls, making two complete circles of the field at Huffman Prairie in 2 minutes and 45 seconds; Dayton, Ohio. Ohio Dayton, 1905. [Sept. 7] Photograph. https://www.loc.gov/item/2001696567/.

Historic Photographs

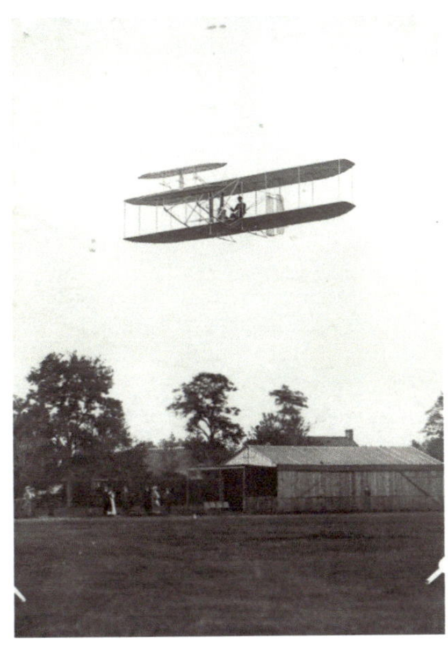

Le Mans Racetrack
Wright State University Library, Special Collections and Archives, MS-1: Wright Brothers Collection, Item identifier number MS1_17_2_6, 8/1/1908, Wilbur Wright circling the field at Les Hunaud-ieres race course near Le Mans, August, 1908.

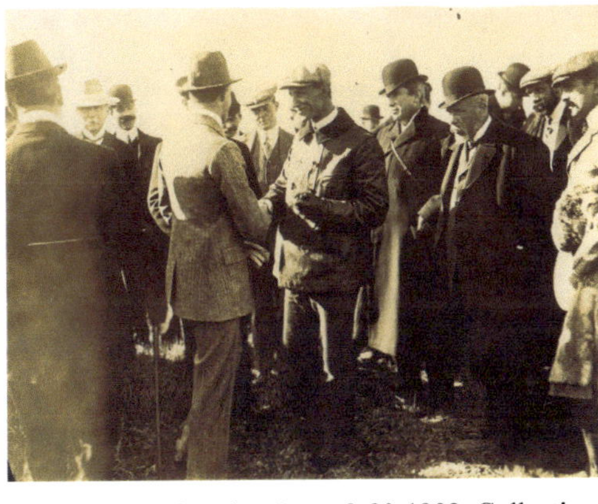

**King
Alfonso XIII
Of Spain**
Wright State University Library, Special Collections and Archives, Item Identifier Number: ms1_18_1_16, Creation Date: 2-20-1909, Collection: MS-1: Wright Brothers Collection, Description: Wilbur Wright smiles as he shakes hands with Alfonso XIII, King of Spain. A group of specta-tors, including student pilot Paul Tissandier (far right, surround Wilbur and the King, Publisher Repository, Special Collections and Archives; Wright State University Libraries, Digital Publisher, Digital Services Department; Wright State University Libraries

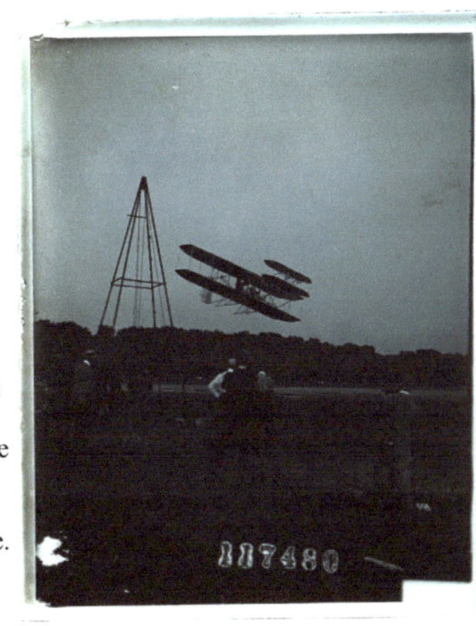

**Speed Test
At Fort
Myer**
National Museum of American History, The Wright aeroplane in flight, Fort Myer. Active no. 10577: non-stereo interpositive. Underwood & Under-wood Glass Stereograph Collection/Series 3/RSN Number 27633-27740/NMAH-AC0143-0027644

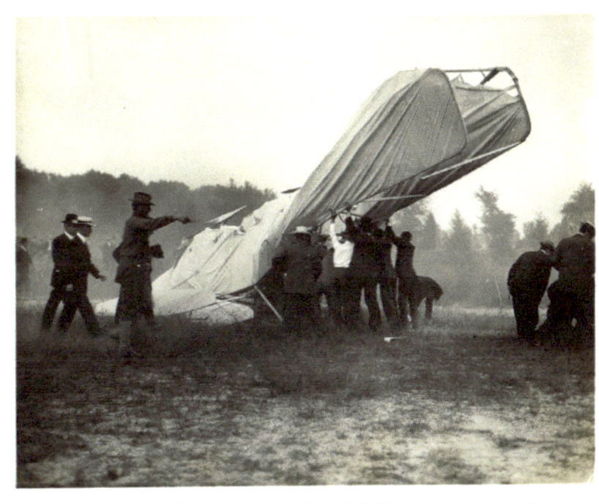

The Crash
National Air and Space Museum, Archives Division, MRC 322, Wash-ington, DC, 20560, Carl H. Claudy Photography Collection, NASM 95-8453, glass negative, 9/17/1908, Four minutes later {Sept 17, 1908} [print #43] Four minutes after start, Sept 17, plane crashed. Soldier pointing is Sergt. Downey, S.C. Man behind, straw hat, Major Squier. Disaster. Orville is still in the wreckage. The little group at right is working over Selfridge. Aftermath of the crash of Orville Wright and Lt. Thomas O. Selfridge in the Wright Type A Military Flyer at Fort Myer, Virginia, September 17, 1908, during US Army flight trials.

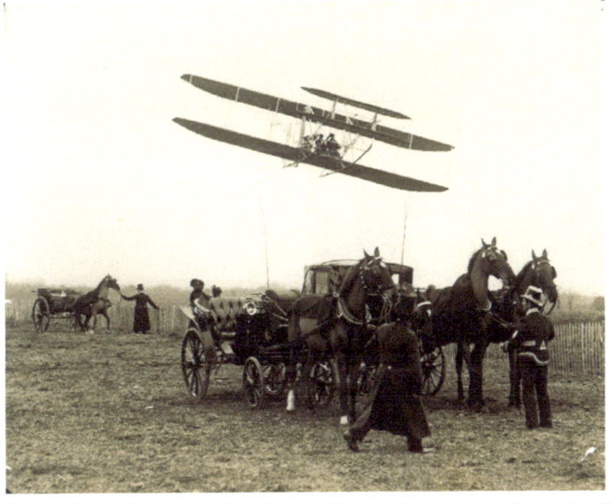

**Horse
Drawn Carriages**
Wright State University, Item Identifier Number: ms1_18_5_8, Creation Date 2-1-1909, Collection MS-1: Wright Brothers Collection, Description: Wilbur Wright in flight with a passenger in the Wright Model A Flyer. The Flyer is flying over horse drawn carriages. This photograph was taken between 02/1909 and 03/1909. Publisher Repository: Special Collections and Archives; Wright State University Libraries, Digital Services Department; Wright State University Libraries.

Historic Photographs

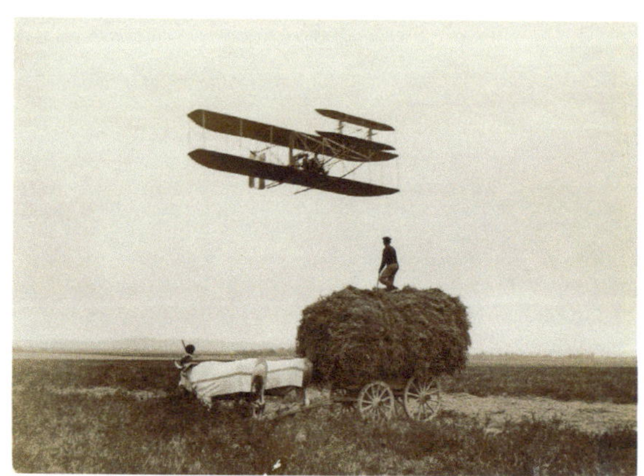

The Infinite Highway Of The Air

Wright State, Item Identifier Number: ms1_18_5_11, Creation Date 2-1-1909, Collection MS-1: Wright Brothers Collection, Description: Wilbur Wright flying the Wright Model A Flyer over ox drawn carts filled with hay. This photograph was taken between 02/1909 and 03/1909. Publisher Repository, Special Collections and Archives; Wright State University Libraries, Digital Services Department; Wright State University Libraries.

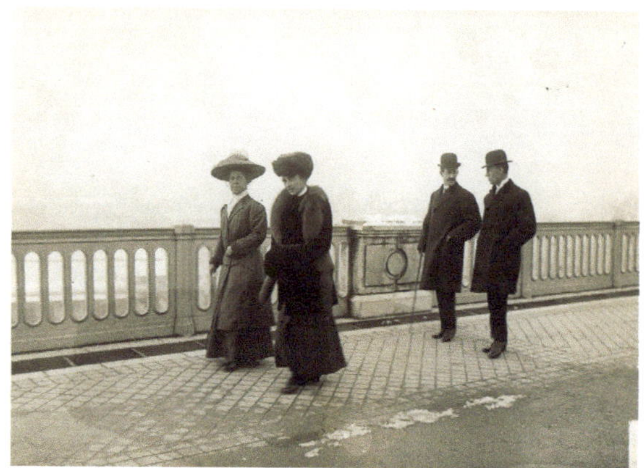

Walking The Boulevard des Pyrénées

Wright State Univ, Item Identifier Number: ms1_18_3_19, Creation Date: 2-1-1909, Collection: MS-1: Wright Brothers Collection, Description: Katharine Wright, Countess Cordelia de Lambert, Orville Wright, and Wilbur Wright walking in Pau, Publisher Repository, Special Collections and Archives; Wright State University Libraries, Digital Services Department; Wright State University Libraries.

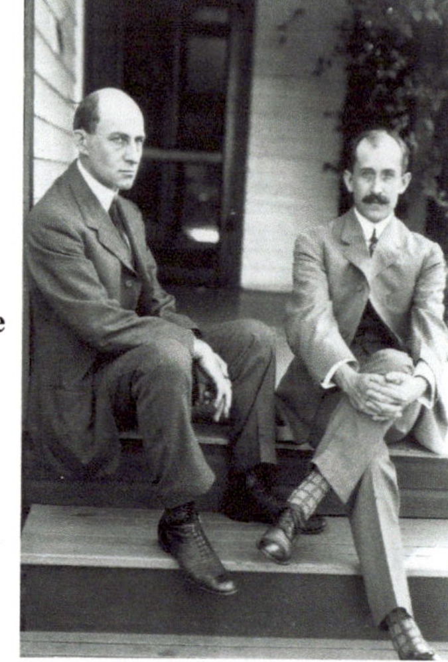

Greenfield Village

Wright State University, Item Identifier Number: ms1_21_1_19, Creation Date:6-1-1909, Collection: MS-1: Wright Brothers Collection, Description: Wilbur (left) and Orville (right) Wright seated on the front porch of their Hawthorne Street home, Publisher Repository: Special Collections and Archives; Wright State University Libraries, Digital Publisher: Digital Services

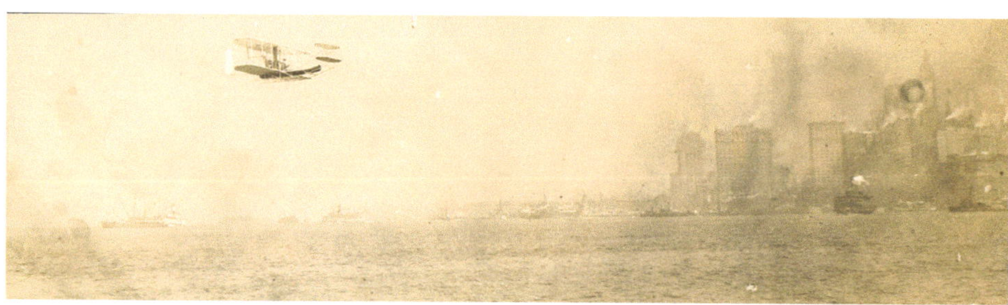

A New Kind Of Gull In New York Harbor

Wright State University, Item Identifier Number: ms1_19_2_6, Creation Date: 10-4-1909, Collection: MS-1: Wright Brothers Collection, Description: Wilbur Wright flying the Wright Model A Flyer up the Hudson River at the Hudson-Fulton Celebration. New York City is in the background, Publisher Repository: Special Collections and Archives; Wright State University Libraries, Publisher: Digital Services Department; Wright State University Libraries.

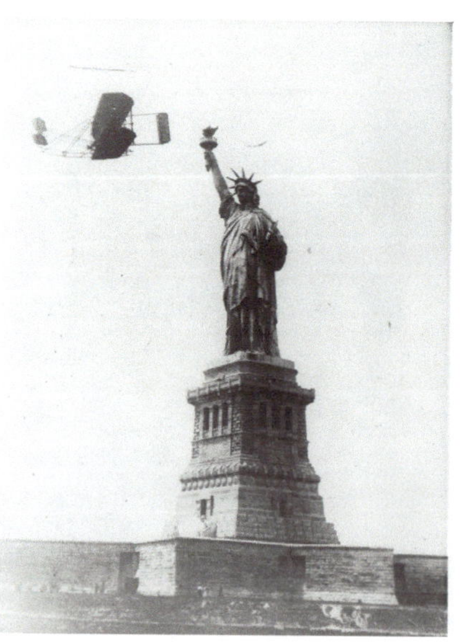

Lady Liberty

Smithsonian National Air and Space Museum, Archives Division, MRC 322, Washington, DC, 20560, SI-85-6235, 9/29/1909, Wilbur Wright flies a Wright Type A by the Statue of Liberty during the Hudson-Fulton Celebration on September 29, 1909.

Historic Photographs

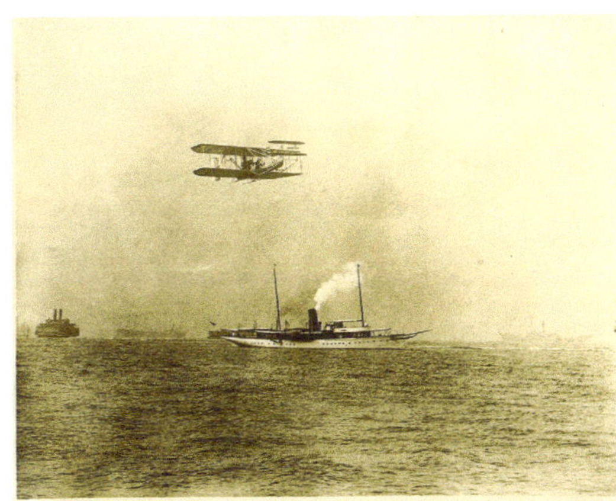

The Hudson-Fulton Celebration
National Air and Space Museum, Archives Division, MRC 322, Washington, DC, 20560, NASM-9A08511, 10/4/1909, "Wilbur Wright during Hudson-Fulton Celebration, New York Oct. 1909." Distant one-half right front view of Wright Type A, piloted by Wilbur Wright, in flight over the steam yacht "Viking" (owned by New York banker George F. Baker, Jr.) in New York Harbor during the Hudson-Fulton Celebration, New York City, October 4, 1909. The biplane has been fitted with a canoe affixed to the underside to provide emergency flotation. Other ships, including a large ferry at left, can be dimly seen in background.

Family
Library of Congress, Wright, Wilbur, and Orville Wright, photographer. Group picture of Orville Wright, Bishop Milton Wright, Katharine Wright, Earl N. Findley, nephew Horace Wright, John R. McMahon, and Pliny Williamson, all seated on the lawn of Orville's home, Hawthorn Hill; Dayton, Ohio. Ohio Dayton, 1915. Photograph. https://www.loc.gov/item/2001696683/.

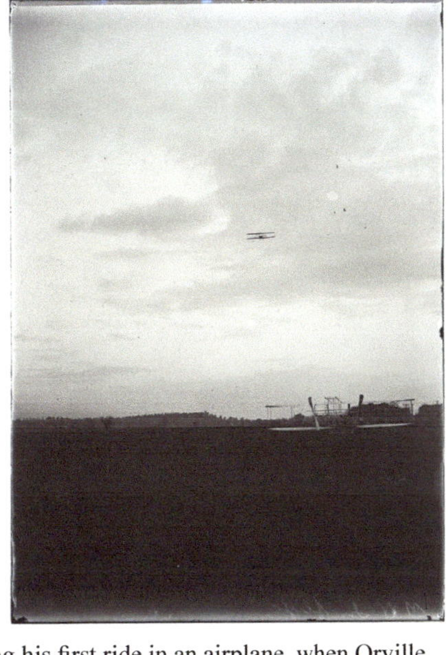

Bishop's 1st Flight
Library of Congress, Wright, Wilbur, and Orville Wright, photographer. Distant view of Bishop Milton Wright during his first ride in an airplane, when Orville attained an altitude of 350 feet at Simms Station, Dayton, Ohio. Ohio Dayton, 1910. Photograph. https://www.loc.gov/item/2001696643/.

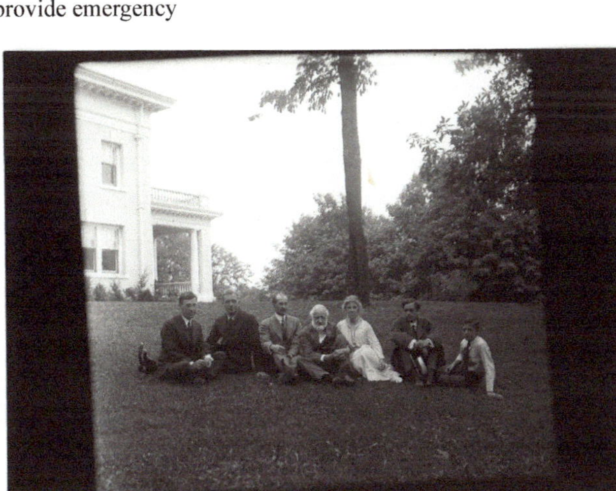

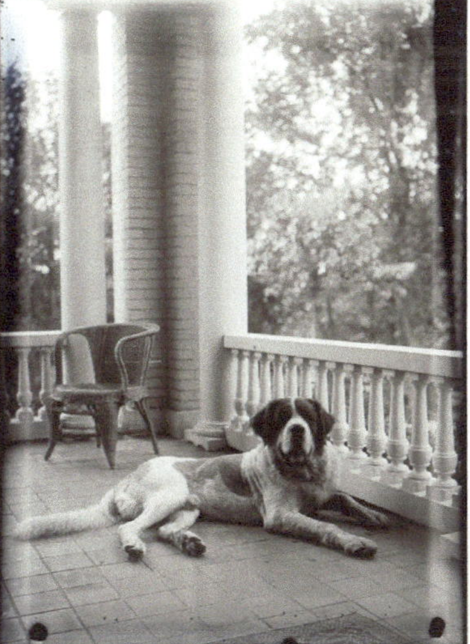

Test Flight On The Miami River
Wright State Univ, Item Identifier Number: ms1_20_2_14, Creation Date: 5-1-1913, Collection: MS-1: Wright Brothers Collection, Description: Wright Model CH Flyer fitted with twin, multi-step pontoons in flight over the Miami River. Two men sitting in a boat on the river are observing the flight. This photographic print is attributed to William Preston Mayfield, Dayton Daily News, Dayton, Ohio, Publisher Repository: Special Collections and Archives; Wright State University Libraries, Digital Publisher: Digital Services Department; Wright State University Libraries.

Scipio
Library of Congress, Wright, Wilbur, and Orville Wright, photographer. Scipio, a St. Bernard dog Orville acquired in March. Ohio Dayton, [Between 1917 and 1928] Photograph. https://www.loc.gov/item/2001696664/.

Historic Quotations

The Bike Shop

"History was being made in their bicycle shop and in their home, but the making was so obscured by the common place I did not recognize it until many years later."
Milton Wright

Milton Wright, nephew of Wilbur and Orville, 1904, Miller, Ivonette Wright, Wright Reminiscences, Privately Published, 1978, page 68, speech by Milton Wright at The Smithsonian museum, presenting the Kitty Hawk Aeroplane in 1948. McCullough, David G. The Wright Brothers. Simon & Schuster, 2016, page 113.

7 Hawthorn Street

"If I were to give a young man advice as to how he might succeed in life, I would say to him, pick out a good Father and Mother and begin life in Ohio."
Wilber Wright

Remarks given by Wilbur Wright at the Twenty-fourth Annual Banquet of the Ohio Society of New York on January 10, 1910, Reports of Proceedings, 1910, New York: Ohio Society of New York, 1910, 93– 138, as cited in Jakab and Young, eds., The Published Writings of Wilbur and Orville Wright, p. 35, footnote 1. McCullough, David G. The Wright Brothers. Simon & Schuster, 2016, page 5.

Kite Flying

"We tried it with the tail in front, behind and every other way. When we got through, Will was so mixed up he couldn't even theorize. It has been with considerable effort that I succeeded in keeping him in the flying business at all."
Orville Wright

From a letter Orville wrote back home to Katharine in 1900, "Wright Brothers Aeroplane Company The Story of Orville and Wilbur Wright, the Invention of the Airplane, and Man's First Flights.", www.wright-brothers.org/. "Wright Stories: Wright Brothers: Inventing The Airplane: History of Flight Kitty Hawk: Wright Contemporaries: Military Airplane." Wright Stories Wright Brothers Inventing The Airplane History of Flight Kitty Hawk Wright Contemporaries Military Airplane RSS, wrightstories.com/.

View From The Top Of Big Kill Devil Hill

"For some years I have been afflicted with the belief that flight is possible to man. My disease has increased in severity and I feel that it will soon cost me an increased amount of money if not my life. I have been trying to arrange my affairs in such a way that I can devote my entire time for a few months to experiment in this field."
Wilbur Wright

Wilbur Wright letter to Octave Chanute on May 13, 1900, Wilbur and Orville Wright Papers at the Library of Congress: Digital Collections: Library of Congress. The Library of Congress, www.loc.gov/collections/wilbur-and-orville-wright-papers/. "The Wright Story/Inventing the Airplane/Letter to Chanute." Afflicted with the Belief, wrightbros. org/History_Wing/Wright_Story/Inventing_the_Airplane/Kitty_Hawk/Afflicted.htm

Gliding Off Of Big Kill Devil Hill

"It has occurred to me that you would get still flatter glides by making sure that the center of gravity coincides exactly with the center of pressure, and possible by decreasing the angle of your rudder to 4 degrees. Please take plenty of snapshots."
Octave Chanute

Letter from Octave Chanute to Wilbur Wright, August 19, 1901, Wilbur and Orville Wright Papers at the Library of Congress: Digital Collections: Library of Congress. General Correspondence: Chanute, Octave. 1901 Manuscript/Mixed Material. Retrieved from the Library of Congress, <www.loc.gov/item/wright002401/>.

Waiting To Fly

"Wind of 6 to 7 meters blowing from west and northwest in morning. We completed repairs by noon and got the machine out on the tracks in front of the building ready for a trial from the level. The wind was gradually dying and by the time we were ready was blowing only about 4 to 5 meters per sec. After waiting several hours to see whether it would breeze up again, we took the machine back in."
Orville Wright

Orville Wright's diary entry on December 16, 1903, Wilbur and Orville Wright Papers at the Library of Congress: Digital Collections: Library of Congress. Diaries and Notebooks:, Orville Wright. 1903. Manuscript/Mixed Material. Retrieved from the Library of Congress, <www.loc.gov/item/wright002238/>.

Historic Quotations

First Flight

"I got on the machine at 10:35 for the first trial. The wind, according to our anemometers at this time, was blowing a little over 20 miles. The machine lifted from the track just as it was entering the fourth rail. A sudden dart when a little over a hundred feet from the end of the track or a little over 120 feet from the point at which it rose into the air, ended the flight."
Orville Wright

Orville Wright's diary entry on December 17, 1903, Wilbur and Orville Wright Papers at the Library of Congress: Digital Collections: Library of Congress. Diaries and Note-books:, Orville Wright. 1903. Manuscript/Mixed Material. Retrieved from the Library of Congress, <www.loc.gov/item/wright002238/>.

Brothers

"From the time we were little children my brother Orville and myself lived together, played together and worked together, and in fact thought together. We talked over our thoughts and our aspirations so that nearly everything that was done in our lives has been the results of conversations, suggestions and discussions between us."
Wilbur Wright

Quote from Wilbur Wright in 1912. "Wright Brothers Aeroplane Company: The Story of Orville and Wilbur Wright, the Invention of the Airplane, and Man's First Flights." Brothers.org, www.wright-brothers.org/, National Air and Space Museum, airandspace. si.edu/. Miller, Ivonette Wright, Wright Reminiscences, Privately Published, 1978, page 122, written by Wilbur in April 1912.

Flight #85

"When you know, after the first few minutes, that the whole mechanism is working perfectly, the sensation is so keenly delightful as to be almost beyond description. Nobody who has not yet experienced it for himself can realize it."
Wilbur Wright

Wilber learning to fly at Huffman Prairie from an interview in the New York Herald, November 25, 1906, McCullough, David G. The Wright Brothers. Simon & Schuster, page 126.

Flight #41

"The ground was a perfect blur, but as the plane rose higher the objects below became clearer. At the height of a hundred feet you feel almost no motion at all, except the wind that strikes your face. If you did not take the precaution to fasten your hat before starting, you have probably lost it by now."
Orville Wright

"The Wright Brothers' Aeroplane," Century Magazine, September 1908; Jakab and Young, eds., The Published Writings of Wilbur and Orville Wright, 32, McCullough, David G. The Wright Brothers. Simon & Schuster, 2016, page 126.

Traveling With The Wind

"You find yourself facing toward the point from which you started. The objects on the ground now seem to be moving at a much higher speed, though you perceive no changes in the pressure of the wind on your face. You know that you are traveling with the wind."
Orville Wright

Century Magazine, September 1908, https://digitalcollections.nypl.org/collections/ century-company-records/, Aviation magazine, December 17, 1923, https://archive. aviationweek.com/issue/19231217, McCullough, David G. The Wright Brothers. Simon & Schuster, 2016, page 126.

Amos Root

"When it turned that circle, and came near the starting point, I was right in front of it, and I said then, and I believe still, it was the greatest sight of my life. Imagine a locomotive that left its tracks and climbed up in the air right toward you. A loco- motive without any wheels but with wings instead, spread 20 feet each way, coming right towards you with a tremendous flap of its propeller and you have something like what I saw. I tell you friends the sensation that one feels is something hard to describe."
Amos Root

Amos Root witnessing a full 180 degree turn September 20, 1904. Published in Glean-ings in Bee Culture, page 38, January 1, 1905, https://www.loc.gov/item/wright002982/, McCullough, David G. The Wright Brothers. Simon & Schuster, 2016, page 120.

Historic Quotations

The Infinite Highway Of The Air

"The desire to fly is an idea handed down to us by our ancestors who, in their grueling travels across trackless lands in prehistoric times, looked enviously on the birds soaring freely through space, at full speed, above all obstacles, on the infinite highway of the air."
Wilbur Wright

Wilbur Wright spoke to members of the Aéro-Club de France November 5, 1908, New York Times, November 6, 1908; Paris Herald, November 6, 1908; L'Aérophile, November 6, 1908, McCullough, David G. The Wright Brothers. Simon & Schuster, 2016, page 208.

Horse Drawn Carriages

"Scarcely ten years ago, all hope had almost been abandoned; even the most convinced had become doubtful, and I confess that in 1901 I said to my brother Orville that men would not fly for fifty years. Two years later, we ourselves were making flights."
Wilbur Wright

Wilbur Wright spoke to members of the Aéro-Club de France November 5, 1908, New York Times, November 6, 1908; Paris Herald, November 6, 1908; L'Aérophile, November 6, 1908, McCullough, David G. The Wright Brothers. Simon & Schuster, 2016, page 208.

Walking the Boulevard des Pyrénées

"Every time we make a move, the people on the street stop and stare at us. We have our pictures taken every two minutes."
Katharine Wright

Katharine Wright, Katherine to Lorin Wright, January 24, 1909, Wilbur and Orville Wright Papers at the Library of Congress: Digital Collections: Library of Congress. The Library of Congress, www.loc.gov/collections/wilbur-and-orville-wright-papers/about-this-collection/, McCullough, David G. The Wright Brothers. Simon & Schuster, 2016, page 213.

Le Mans Racetrack

"It is not really necessary to look too far into the future. We have seen enough already to see that it will be magnificent. Only let us hurry and open the roads."
Wilbur Wright

Wilbur Wright spoke to members of the Aéro-Club de France November 5, 1908, New York Times, November 6, 1908; Paris Herald, November 6, 1908; L'Aérophile, November 6, 1908, McCullough, David G. The Wright Brothers. Simon & Schuster, 2016, page 208.

King Alfonso XIII Of Spain

"Princes and millionaires are as thick as fleas on the Flyer"
Wilbur Wright

From a letter Wilbur wrote to Katharine in Dayton. Papers of Wilbur and Orville Wright Papers at the Library of Congress, McCullough, David G. The Wright Brothers. Simon & Schuster, page 205.

Speed Test At Fort Myer

"He flew 67 minutes in a 16 mile wind, handling his plane like a chauffeur, and road the wind as deliberately as if he were passing over a solid macadam road. Nothing I have ever seen is comparable in action to this gliding bird."
Gutzon Borglum

Crouch, Tom D. Wings: a History of Aviation from Kites to the Space Age. Norton, 2004, page 6, McCullough, David G. The Wright Brothers. Simon & Schuster, 2016, page 185.

The Crash

"As the aeroplane dashed off the rising track, Lieut. Selfridge waved his hand gayly to a group of army officers and newspaper men and threw back some laughing remarks that were drowned out by the whirl of the propellors. As he swept around Selfridge evidently was enjoying himself thoroughly. It was on the fourth and final lap that the propellor blade broke. The aeroplane took a short and sharp dive and crashed into the field. Dust arose in a yellow pall over the great white man-made bird that had dashed to its death."
New York Times

September 18, 1908, New York Times archives, https://timesmachine.nytimes.com/timesmachine/1908/09/18/issue.html.

Historic Quotations

Greenfield Village

"Wilbur and Orville were among the blessed few who combined mechanical ability with intelligence in about equal amounts. One man with this dual gift is exceptional. Two such men whose lives and fortunes are closely linked can raise this combination of qualities to a point where their combined talents are akin to genius."
Fred Howard

Howard, Fred. Wilbur and Orville: A Biography of the Wright Brothers. Dover Publications, 1998.

The Hudson-Fulton Celebration

"I went to a height just a little above the ferry boats until I reached the battleships. I passed so close to the funnels that I could smell the smoke from them"
New York Journal

Wilber interviewed by the New York Journal October 4, 1909 historic archives, https://www.loc.gov/collections/new-york-journal/, McCullough, David G. The Wright Brothers. Simon & Schuster, 2016, p 245.

Lady Liberty

"In the air Wright seemed to pause for a moment to pay the homage of an American aviator to the lady who attests his country's destinies. Then suddenly turning eastward with the wind, he sped rapidly over the waves while the harbor crafts shrieked their welcome, and cheering men and women bore witness that our Lady Liberty had been visited by one of her children in a vessel needing only the winds on which to sail."
New York Evening Sun

New York Evening Sun, September 19, 1909 historic archives, https://nyshistoricnewspapers.org.

A New Kind Of Gull In New York Harbor

"Once his great aeroplane, so near the horizon that it seemed one with the ocean gulls among which it flew"
New York Evening Sun

New York Evening Sun September 29, 1909, historic archives, https://nyshistoricnewspapers.org and Governors Island, www.govisland.com/, "A New Kind Of Gull In New York Harbor" headline Harpers Weekly, October 9, 1909, Patchett, Ann, et al. Harper's Magazine, harpers.org/.

Bishop's First Flight

"Higher, Orville, Higher!"
Milton Wright

Bishop Milton Wright's 1st Flight, Wilbur and Orville Wright Papers at the Library of Congress: Digital Collections: Library of Congress. The Library of Congress, www.loc.gov/collections/wilbur-and-orville-wright-papers/, McCullough, David G. The Wright Brothers. Simon & Schuster, 2016, p 253.

Test Flight On The Miami River

"Isn't it astonishing that all these secrets have been preserved for so many years just so we could discover them!"
Orville Wright

Letter to George A. Spratt June 7, 1908, friend of Orville Wright, Wright, Wilbur, et al. Miracle at Kitty Hawk: Unpublished Letters of the Wright Brothers. Atlantic Monthly Co., 1950, Fred C. Kelly, Wright Brothers official biographer.

Family

"But it isn't true to say we had no special advantage, the greatest thing in our favor was growing up in a family where there was always much encouragement to intellectual curiosity."
Orville Wright

Fred C. Kelly, "Interview with Orville Wright", March 31, 1934; Jakab and Young, eds., The Published Writings of Wilbur and Orville Wright, P 83, McCullough, David G. The Wright Brothers. Simon & Schuster, 2016, p 18.

Scipio

"When Orville Wright dies in 1948 they only found one photograph in his wallet. It wasn't of the first flight, or of his sister or parents or of he and his brother. It was a photograph of his dog Scipio who had died 25 years earlier."
Steven Wright

Story told by Steven Wright, Wright Brothers great-grandnephew at Hawthorn Hill open house tour, WYSO Public Radio, local and statewide news, https://www.wyso.org/news. Miller, Ivonette Wright, Wright Reminiscences, Privately Published, 1978, page 61.

Photographic Process

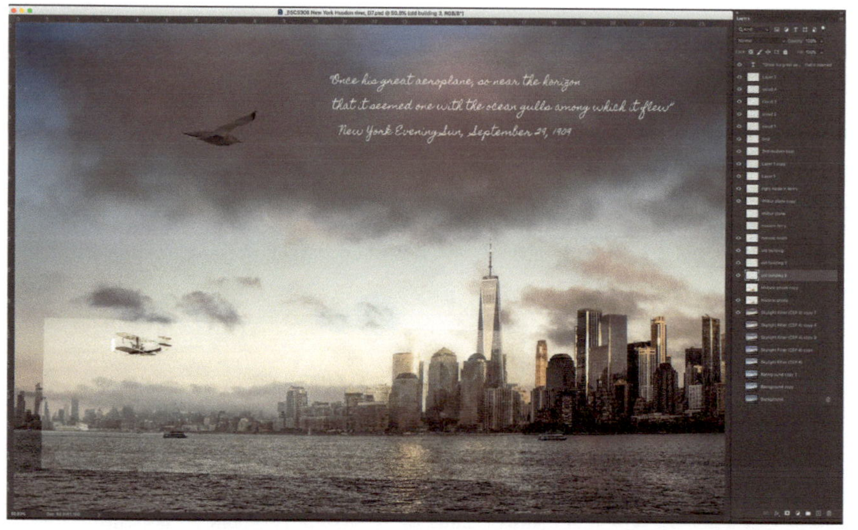

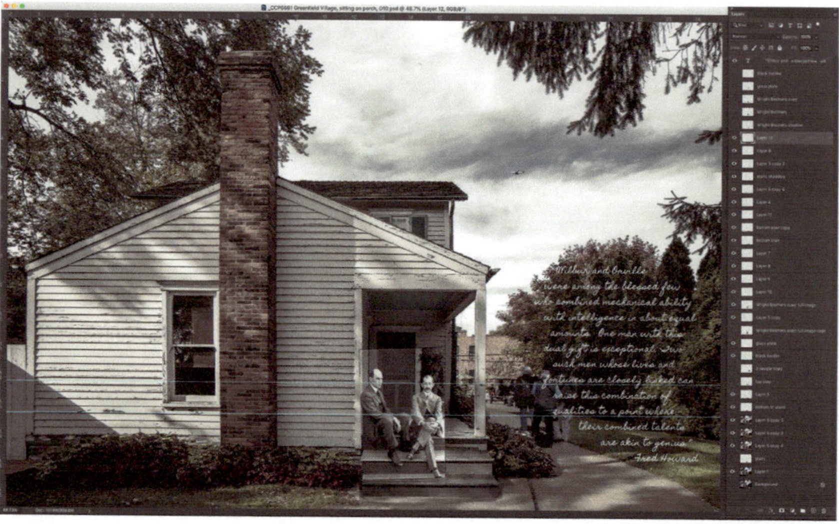

As a photographer and artist who uses PhotoShop to create my images, this computer program allows me to complete my vision. In my series *Wright Brothers: Then and Now*, I use numerous layers to create my photographic collages. Here are five examples showing the number of layers involved in creating the finished photographs.

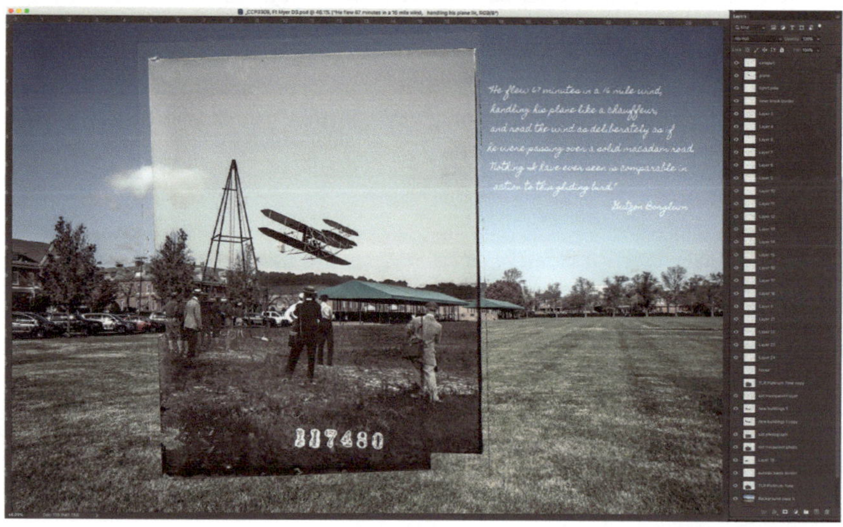

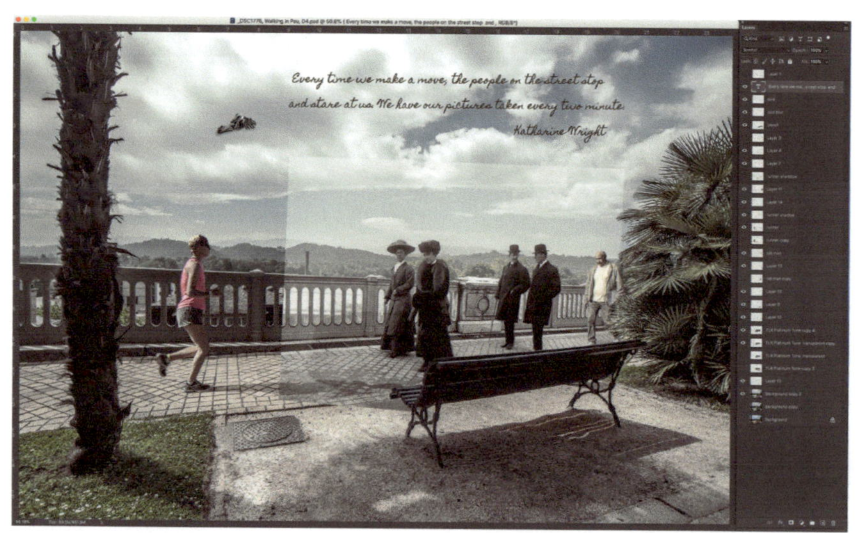

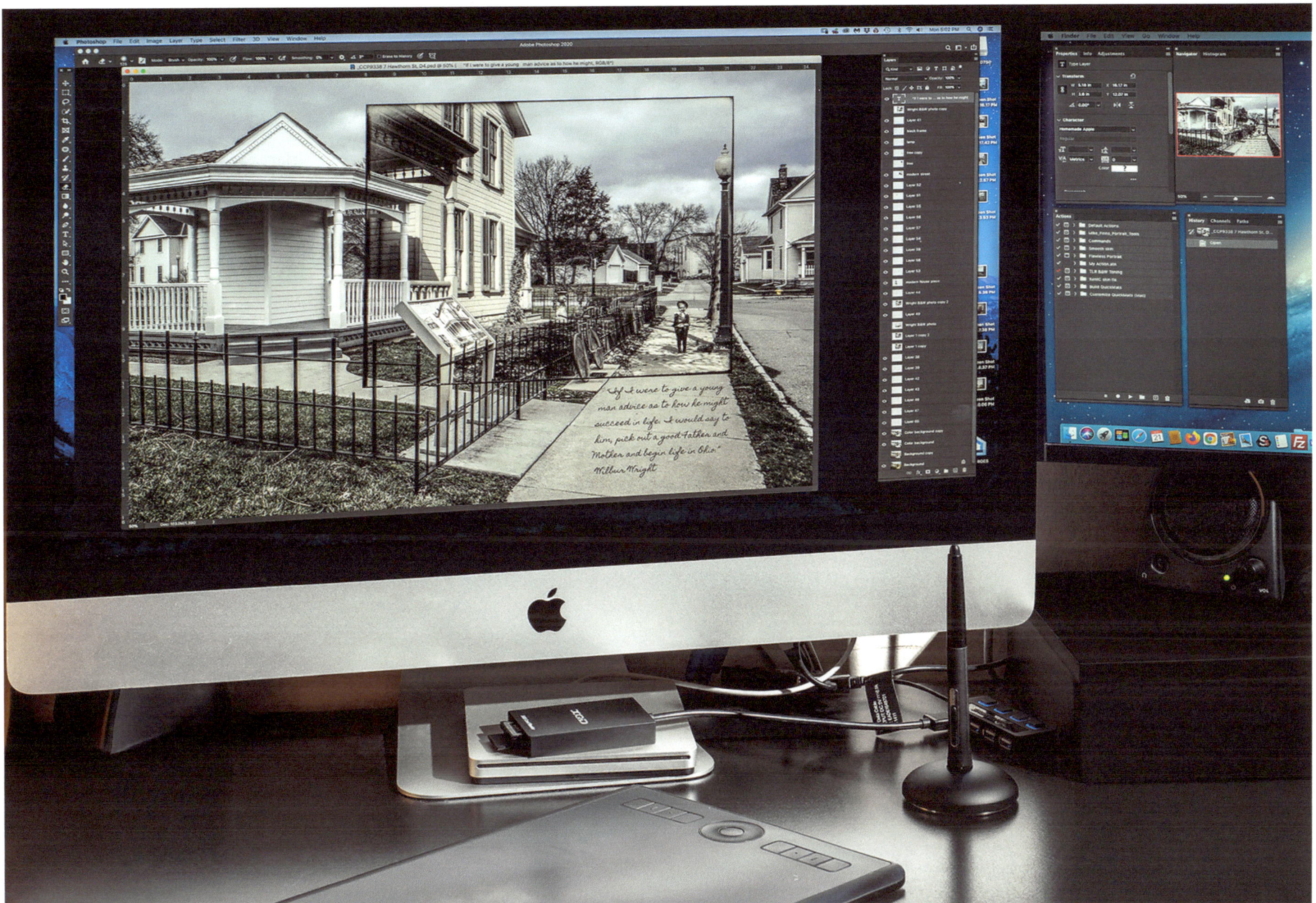

Acknowledgements

To Maria, my wife of soon to be 38 years:
Thank you for all your encouragement, your honest critiques, and being my go-to editor.
None of this would be possible without you.

To Tom Rubens and all my BNI Miracle Morning mastermind friends:
You helped get this project started.

To Eva Buttacavoli at The Contemporary Dayton:
Thank you for helping me see my early vision.

To Dawne Dewey and Bill Stolz at the Wright State University Library Special Collections department:
Thanks for your help with my research and with getting me the digital photographs from your archives.

Thank you to the people of Montgomery County, Ohio, for the Artist Opportunity Grant administered by Culture Works,
which allowed me to travel to France and create photographs for this project.

Thanks to my friend from France, Paul Mirat, my Wright Brothers tour guide in Pau.

Thanks to Sarah Guthrie, my artist business coach with The Abundant Artist.
You kept me on track and helped me move this project forward.

Thanks to Rick Spencer, The WordSmith.
You helped me tell my story.

Thank you to Robert Brocke for your graphic design of this book
and an extra set of eyes on my photographs.

And thanks to David McCullough for writing such an inspirational book about the Wright Brothers.

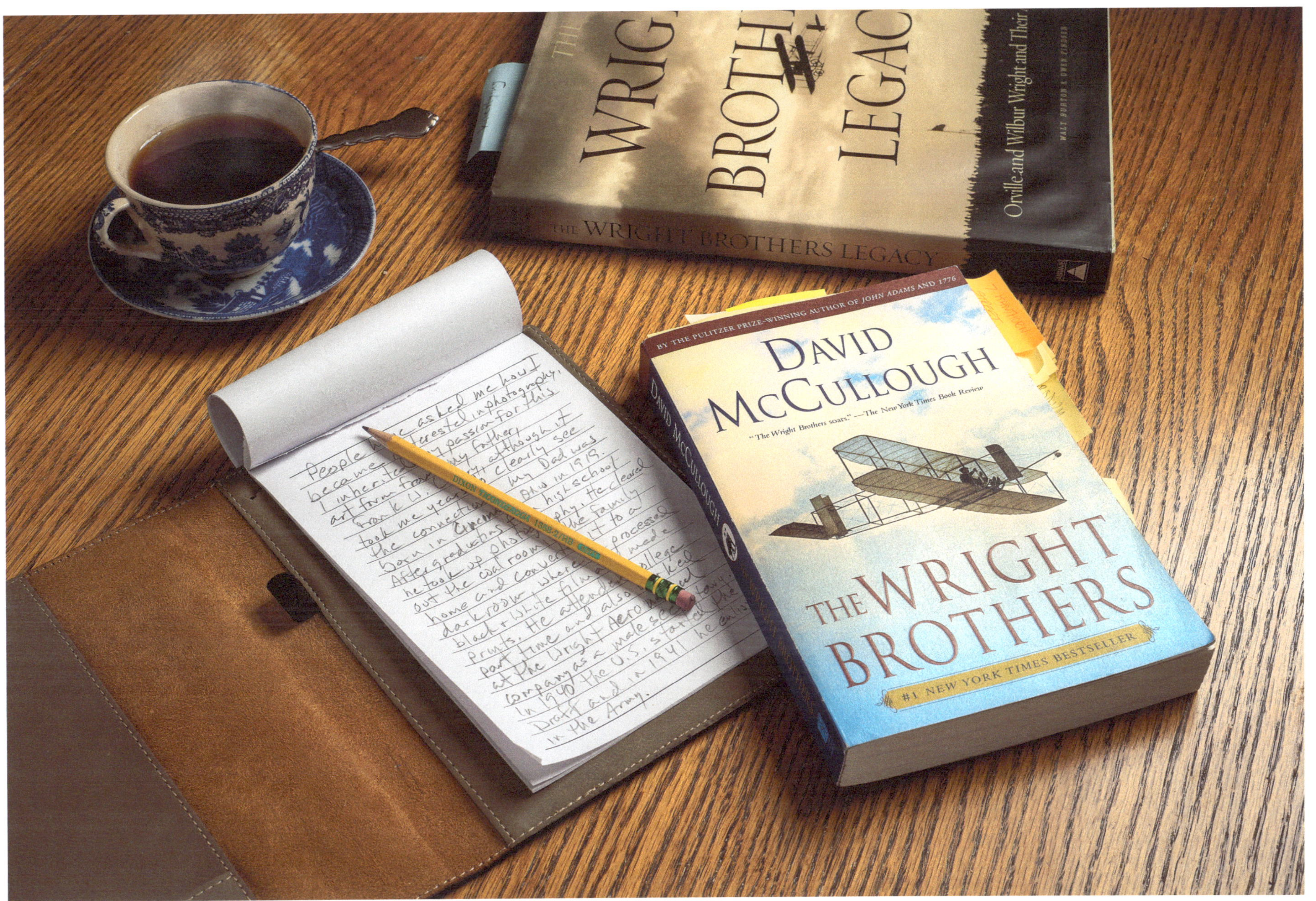

THE WRIGHT BROTHERS LEGACY
Orville and Wilbur Wright and Their
WALT BURTON & OWEN FINDSEN
THE WRIGHT BROTHERS LEGACY
BY THE PULITZER PRIZE-WINNING AUTHOR OF JOHN ADAMS AND 1776
DAVID McCULLOUGH
"The Wright Brothers soars." —The New York Times Book Review
DAVID McCULLOUGH
THE WRIGHT BROTHERS
#1 NEW YORK TIMES BESTSELLER

Source Notes

The Bike Shop: Wright Brothers fourth bike shop…Visitnaha.com, visitnaha.com/aviation_site/national-park-service-visitor-center-aviation-parachute-museum/wright-cycle-company/.

7 Hawthorn Street: Orville was interested in photography…"Photography and the Wright Brothers: Articles and Essays: Wilbur and Orville Wright Papers at the Library of Congress: Digital Collections: Library of Congress." The Library of Congress, www.loc.gov/collections/wilbur-and-orville-wright-papers/articles-and-essays/photography-and-the-wright-brothers/.

Kite Flying: "I chose Kitty Hawk because…", "Kitty Hawk (U.S. National Park Service)." National Parks Service, U.S. Department of the Interior, www.nps.gov/places/kittyhawk.htm.

View From The Top Of Big Kill Devil Hill: Octave Chanute was an American civil engineer and aviation pioneer… Wright Brothers, The Recovered Legacy: Octave Chanute, memory.loc.gov/master/ipo/qcdata/qcdata/wrightold/wb005.html.

Gliding Off Of Big Kill Devil Hill: The monument and National Park are actually in Kill Devil Hills, North Carolina… "Kill Devil Hills, North Carolina." Wikipedia, Wikimedia Foundation, 19 Jan. 2021, en.wikipedia.org/wiki/Kill_Devil_Hills,_North_Carolina. "Please take plenty of snapshots" "Octave Chanute Papers: Special Correspondence--Wright Brothers, 1900." The Library of Congress, www.loc.gov/item/wright002804/.

Waiting To Fly: The Wright Brothers had been working hard for the past month to finish the flying machine, but bad weather and mechanical problems held them up… McCullough, David G. The Wright Brothers. Simon & Schuster, 2016, page 99 – 102, The quote used in this photograph is from Orville's diary on December 16th…"Diaries and Notebooks: 1903, Orville Wright." The Library of Congress, www.loc.gov/item/wright002238/.

First Flight: John T. Daniels took the famous photograph of the Wright Brothers' first flight… McCullough, David G. The Wright Brothers. Simon & Schuster, 2016, page 105.

Brothers: The brothers made 105 flights at Huffman Prairie during that year, totaling 49 minutes in the air… "Huffman Prairie Flying Field." National Parks Service, U.S. Department of the Interior, www.nps.gov/daav/learn/historyculture/huffman-prairie-flying-field.htm.

Amos Root: I could hear Amos Root's words in my head… McCullough, David G. The Wright Brothers. Simon & Schuster, 2016, page 117 – 120.

Flight #85: On that day, Orville flew a distance of 1,760 feet in 40 seconds or about 30 miles per hour… Wright, Wilbur, and Orville Wright. "[Flight 85: Orville in Flight over Treetops, Covering a Distance of Approximately 1,760 Feet in 40 1/5 Seconds; Huffman Prairie, Dayton, Ohio]." Home, 1 Jan. 1970, www.loc.gov/pictures/item/2001696497/.

Flight #41: The Wrights had to navigate around a honey locust tree in the middle of the field while making test flights… McCullough, David G. The Wright Brothers. Simon & Schuster, 2016, page 125.

Traveling With The Wind: I paired the image I made that morning with the historical photograph of Flight #23 taken at Huffman Prairie on September 7, 1905. Orville was at the controls, making two complete circles of the field in 2 minutes and 45 seconds. I could imagine Orville using the tree as a target to circle while perfecting flight during the 1904 and 1905 trials. "[Flight 23: Front View of the Machine in Flight to the Right, Orville at the Controls, Making Two Complete Circles of the Field at Huffman Prairie in 2 Minutes and 45 Seconds; Dayton, Ohio]." The Library of Congress, www.loc.gov/item/2001696568/. McCullough, David G. The Wright Brothers. Simon & Schuster, 2016, page 125

Le Mans Racetrack: Wilbur Wright gave the first public flying machine demonstration on August 8, 1908, at Les Hunaudières racetrack in Le Mans, France. In 1923, just 15 years after his flight, this same racetrack would become part of the world-famous 24 Hours of Le Mans automobile race. McCullough, David G. The Wright Brothers. Simon & Schuster, 2016, page 167, "24 Hours of Le Mans." Wikipedia, Wikimedia Foundation, 15 Jan. 2021, en.wikipedia.org/wiki/24_Hours_of_Le_Mans. … "wood pieces from the airplane Kitty Hawk went to the moon…"Plaque, Wright Brothers 1903 and Apollo 11 Flights." National Air and Space Museum, airandspace.si.edu/collection-objects/plaque-wright-brothers-1903-and-apollo-11-flights/nasm_A19721288000.

King Alfonso XIII Of Spain: Taking long walks in Paris and visiting the Louvre Museum were among his favorite things to do. The inventor slept in the hanger with his airplane instead of staying in the local hotel. McCullough, David G. The Wright Brothers. Simon & Schuster, 2016, page 140, 166.

Speed Tetst At Fort Myer: On Thursday, September 3, 1908, Orville made his first test flight for the US Army at Fort Myer…. McCullough, David G. The Wright Brothers. Simon & Schuster, 2016, pages 181-185.

The Crash: Lieutenant Thomas Selfridge was keenly interested in aviation and was the first US Military officer to pilot a modern airplane 100 yards in May 1908… McCullough, David G. The Wright Brothers. Simon & Schuster, 2016, page 191, "Thomas Selfridge." Wikipedia, Wikimedia Foundation, 22 Jan. 2021, en.wikipedia.org/wiki/Thomas_Selfridge.

Source Notes

Horse Drawn Carriages: "Virtually every day but Sunday a steady stream of elegant carriages and automobiles headed out to the flying field..." McCullough, David G. The Wright Brothers. Simon & Schuster, 2016, page 216.

The Infinite Highway Of The Air: Mirat, Paul. "Autrefois Pau L'aviation." Amazon, Atlantica, 2004, www.amazon.com/Autrefois-Pau-laviation/dp/2843947391.

Walking The Boulevard des Pyrénées: France. By this time, the Wrights had become celebrities, and Katharine was enjoying the notoriety. McCullough, David G. The Wright Brothers. Simon & Schuster, 2016, page 216.

Greenfield Village: In 1937, Henry Ford purchased the Wright Brothers' house at 7 Hawthorn Street and had it moved to Greenfield Village in Dearborn, Michigan. "Wright Brothers Home and Cycle Shop, Greenfield Village, 1971." *The Henry Ford,* www.thehenryford.org/collections-and-research/digital-collections/artifact/373441/. Equally important to me was the backyard shed, which was also there. The Wright Brothers used the shed as their darkroom…Wright-Miller, Ivonette. *Uncle Wil & Uncle Orv: Wright Brothers Recollections Recorded by Niece, Ivonette Wright Miller and Nephew, Horance Wright.*

The Hudson-Fulton Celebration: New York and New Jersey had a commemoration of the 300th anniversary of Henry Hudson's discovery of the Hudson River and Robert Fulton's 100th anniversary of the paddle steamer. https://en.wikipedia.org/wiki/Hudson%E2%80%93Fulton_Celebration. Wilbur Wright and Glenn Curtiss, the two most celebrated pilots of the day, gave flight demonstrations on the Hudson River as part of the celebration… McCullough, David G. The Wright Brothers. Simon & Schuster, 2016, page 242-247. Wilbur is flying up the Hudson River past the military battleships and over the yacht Viking, owned by New York banker George Baker, Jr. National Air and Space Museum, Archives Division, MRC 322, Washington, DC, 20560, NASM-9A08511, 10/4/1909, "Wilbur Wright during Hudson-Fulton Celebration, New York Oct. 1909." Distant one-half right front view of Wright Type A, piloted by Wilbur Wright, in flight over the steam yacht "Viking" (owned by New York banker George F. Baker, Jr.) in New York Harbor during the Hudson-Fulton Celebration. "Taking Flight with Lady Liberty." *National Air and Space Museum,* 1 July 2020, airandspace.si.edu/stories/editorial/taking-flight-lady-liberty.

Lady Liberty: I love how the four young ladies…"Taking Flight with Lady Liberty." *National Air and Space Museum,* 1 July 2020, airandspace.si.edu/stories/editorial/taking-flight-lady-liberty.

A New Kind Of Gull In New York Harbor: serendipity: "The occurrence of events by chance in a happy or beneficial way." Google Search, Google, www.google.com/search?client=firefox-b-1-d&sxsrf=ALeKk01sRaRO_Sz3zpZyTCKVE2BK67Jx-6Q%3A1613016140758&ei=TKwkYPDiLcvWtAb09roDQ&q=serendipity%2Bmeaning&oq=ser&gs_lcp=Cgdnd3Mtd2l6EAEYADIECCMQJzIFCC4QkQIyBAgAEEMyBAgAEEMyCgguELEDEIMBEEMyBwgAEBQhwIyBAgAEEMyBQgAELEDMgUIABCxAzIFCC4QsQM6BwgAEEcQsAM6BQgAEJECOgsILhCxAxDHARCjAjoICAAQsQMQgwE6AgguOggILhCxAxCDAToCCCABQsz1YmERg4F5oAnACeA-CAAbUBiAHzA5IBAzEuM5gBAKABAAoBB2d3cy13aXrIAQfAAQE&sclient=gws-wiz. The title of this images…"Taking Flight with Lady Liberty." *National Air and Space Museum,* 1 July 2020, airandspace.si.edu/stories/editorial/taking-flight-lady-liberty.

Bishop's First Flight: McCullough points out that Wilbur and Orville had never flown together, ensuring that the other could continue their work in case of a tragedy… McCullough, David G. The Wright Brothers. Simon & Schuster, 2016, pages 252-253. Library of Congress, Wright, Wilbur, and Orville Wright, photographer. Distant view of Bishop Milton Wright during his first ride in an airplane, when Orville attained an altitude of 350 feet at Simms Station, Dayton, Ohio. Ohio Dayton, 1910. Photograph. https://www.loc.gov/item/2001696643/.

Test Flight On The Miami River: The photograph is attributed to Preston Mayfield of the Dayton Daily Newspaper and was taken on May 1, 1913. Wright State Univ, Item Identifier Number: ms1_20_2_14, Creation Date: 5-1-1913, Collection: MS-1: Wright Brothers Collection, Description: Wright Model CH Flyer fitted with twin, multi-step pontoons in flight over the Miami River. Two men sitting in a boat on the river are observing the flight. This photographic print is attributed to William Preston Mayfield, Dayton Daily News, Dayton, Ohio, Publisher Repository: Special Collections and Archives; Wright State University Libraries, Digital Publisher: Digital Services Department; Wright State University Libraries.

Family: Wilbur Wright contracted typhoid fever and died on May 30, 1912. "Wright Brothers." Wikipedia, Wikimedia Foundation, 8 Feb. 2021, en.wikipedia.org/wiki/Wright_brothers. McCullough, David G. The Wright Brothers. Simon & Schuster, 2016, page 256. It was completed in the spring of 1914, and on April 28, Bishop Wright and Katharine moved in. Orville, on a business trip, would follow two days later. "100 Years Ago Today: Wrights Move into Hawthorn Hill." *Out of the Box,* 23 Apr. 2014, www.libraries.wright.edu/community/outofthebox/2014/04/28/100-years-ago-today-wrights-move-into-hawthorn-hill/.

Scipio: Milton Wright's diary entry of March 10, 1917, states, "Scipio came. He weighs 16 pounds. He is a St. Bernard dog. He is a good-looking puppy." Wright, Milton. "Diaries 1857-1917." CORE Scholar, corescholar.libraries.wright.edu/milton_wright_diaries/42/page 820.

www.ingramcontent.com/pod-product-compliance
Lightning Source LLC
Chambersburg PA
CBHW042032050726
47599CB00006B/874